DICTIONARY
of
AUDIO & HI-FI

by

The Howard W. Sams Editorial Staff

HOWARD W. SAMS & CO., INC.
THE BOBBS-MERRILL CO., INC.
INDIANAPOLIS · KANSAS CITY · NEW YORK

FIRST EDITION

FIRST PRINTING—1975

International Standard Book Number: 0-672-21084-3
Library of Congress Catalog Card Number: 74-79354

PREFACE

As new inventions and technologies are developed, new terminology is often needed so that the ideas and techniques involved may be easily and accurately passed from one person to another. The designer, the manufacturer, and, eventually, the consumer of a new product all need to be able to communicate intelligently with each other about the product. This can be done only when the terms used in such communication have the same meaning to all persons involved. Hence, the need for dictionaries, for that is the principal function of a dictionary such as this one: to establish the meaning of the terms that appear within.

The ideal dictionary would be continually changing to keep pace with the vocabulary of a fast-changing technology. New words are constantly being added to a vocabulary; conversely, old words sometimes fall into disuse and may be dropped, but often they are retained for their historical value. Much care has been exercised to make this dictionary as up to date as possible.

Although this is a dictionary of audio terms, many of the included terms would seem equally appropriate in a dictionary of electronics or a dictionary of physics. This is inevitable because by nature these sciences are closely overlapping. There is no sharp dividing line between them.

The alphabetization of the entries in this book follows standards accepted by prominent lexicographers. All terms of more than one word are treated as if they were one continuous word. Thus, "effective sound pressure" appears between "effective acoustic center" and "effective value." Abbreviations are also treated alphabetically.

When more than one definition is given for a term, each definition is numbered. This does not necessarily imply the same order of importance.

Illustrations are positioned to be near the terms they illustrate and their captions agree exactly or very closely with the entry term.

Entry terms are given in the form in which they would naturally occur when spoken or when used in written text. Thus, "chord organ" appears in that order rather than as "organ, chord."

For ease in quickly locating a specific term, catchwords for the first and last entries on each page are shown at the top of the page.

A

A—Symbol for ampere.

absorption coefficient—Ratio of sound absorbed by a material to total sound received. Thus, if a material absorbed all sound, it would have a coefficient of 1. Since all materials reflect some sound, practical absorption coefficients are always less than 1, or a decimal fraction. Also called acoustic absorptivity.

absorptivity—A measure of the portion of incident radiation or sound energy absorbed by a material.

A-B test—Direct comparison of two sounds by playing first one and then the other. May be done with two tape recorders playing identical tapes, two speakers playing alternately from the same tape recorder, two amplifiers playing alternately through one speaker, etc.

ac—Abbreviation for alternating current.

ac bias—In a tape recorder, the alternating current that is fed to the record head in addition to the signal current. The bias frequency is usually several times higher than the highest signal frequency. The ac bias serves to linearize the recording process.

Material	*Coefficients*					
	125 Hz	*250 Hz*	*500 Hz*	*1000 Hz*	*2000 Hz*	*4000 Hz*
Brick, unglazed	0.03	0.03	0.03	0.04	0.05	0.07
Carpet, heavy, on concrete	0.02	0.06	0.14	0.37	0.60	0.65
Carpet, with latex backing on 40-oz hairfelt of foam rubber	0.08	0.27	0.39	0.34	0.48	0.63
Concrete block, coarse	0.36	0.44	0.31	0.29	0.39	0.25
Light velour, 10 oz per sq-yd in contact with wall	0.03	0.04	0.11	0.17	0.24	0.35
Concrete or terrazo	0.01	0.01	0.015	0.02	0.02	0.02
Wood	0.15	0.11	0.10	0.07	0.06	0.07
Glass, large heavy plate	0.18	0.06	0.04	0.03	0.02	0.02
Glass, ordinary window	0.35	0.25	0.18	0.12	0.07	0.04
Gypsum board, nailed to 2 by 4 studs on 16-inch centers	0.29	0.10	0.05	0.04	0.07	0.09
Plaster, gypsum, or lime, smooth finish on tile or brick	0.013	0.015	0.02	0.03	0.04	0.05
Plywood, ⅜ inch	0.28	0.22	0.17	0.09	0.10	0.11
Air, sabins per 1000-cu. ft.	—	—	—	—	2.3	7.2
Audience, seated in upholstered seats, per sq. ft. of floor area	0.44	0.54	0.60	0.62	0.58	0.50
Wooden pews, occupied, per sq. ft. of floor area	0.57	0.61	0.75	0.86	0.91	0.86
Chairs, metal or wooden, seats unoccupied	0.15	0.19	0.22	0.39	0.38	0.30

Absorption coefficients for different materials.

accompaniment manual—In an organ, the keyboard used for playing the accompaniment to the melody. Also called the lower manual, or great manual.

ac erase head—In magnetic recording, a head to which high-frequency current is supplied to produce the magnetic field necessary for removal of previously recorded information.

acetate—A plastic material used as the base for some magnetic recording tapes.

A-channel—One of two stereo channels (usually the left) of a stereo system.

acoustic—Also acoustical. Pertaining to sound or the science of sound.

acoustic absorption loss—The energy lost by conversion to heat or other forms when sound is absorbed by a medium.

acoustic absorptivity— *See* Absorption Coefficient.

acoustical-electrical transducer—A device designed to transform sound energy into electrical energy and vice versa.

acoustical ohm—A measure of acoustic resistance, reactance, or impedance. One acoustical ohm is equal to a volume velocity of 1 cubic centimeter per second when produced by a sound pressure of 1 microbar.

acoustic capacitance—In a sound medium, a measure of volume displacement per dyne per square centimeter. The unit is centimeter to the fifth power per dyne.

acoustic compliance—1. The measure of volume displacement of a sound medium when subjected to sound waves. 2. That type of acoustic reactance corresponding to capacitive reactance in an electrical circuit.

acoustic delay line—A device that retards signal vibrations by causing them to pass through a solid or a liquid.

acoustic dispersion—The separation of a complex sound wave into its various frequency components because of the different wave velocities of each frequency in a given medium.

acoustic elasticity—1. The compressibility of the air in a speaker enclosure as the cone moves backward. 2. The compressibility of any material through which sound is passed.

acoustic feedback—If positive in phase, also called acoustic regeneration. The mechanical coupling of a portion of the sound waves from the output of an audio-amplifying system to a preceding part or input circuit (such as the microphone) of the system. When excessive, positive acoustic feedback will produce a howling sound in the speaker.

acoustic filter—A sound-absorbing device that suppresses certain audio frequencies while allowing others to pass.

acoustic generator—A transducer such as a speaker, headphones, or a bell, that converts electrical, mechanical, or other forms of energy into sound.

acoustic horn—Also called horn. A tube of varying cross section designed to match the impedance of the sound generator at the horn input to the impedance of the surrounding air at the horn output, or mouth.

acoustic impedance—Total opposition of a medium to sound waves, including both acoustic resistance and acoustic reactance. Numerically equal to the force per unit area on the surface of the medium, divided by the flux through that surface. The unit of acoustic impedance is the acoustic ohm. One unit is equal to a volume velocity of 1 cubic centimeter per second produced by a pressure of 1 microbar.

acoustic inertance—Opposition to movement of a sound medium offered because of the inertia of the effective mass of the medium. Analogous to inductive reactance in an electrical circuit. Measured in acoustic ohms.

acoustic labyrinth—Special speaker enclosure having partitions and passages to prevent cavity resonance and to reinforce bass response.

acoustic lens—1. An array of mechanical structures that refract sound waves in much the same way that an optical lens refracts light waves. 2. A device that produces convergence or divergence of moving sound waves. When used with a speaker, the acoustic lens widens the beam of the higher-frequency sound waves.

acoustic ohm—The unit of acoustic resistance, reactance, or impedance. One acoustic ohm is involved when a sound pressure of 1 dyne per square centimeter produces a volume velocity of 1 cubic centimeter per second.

acoustic pickup—In nonelectrical phonographs, the method of reproducing the material on a record by linking the needle directly to a flexible diaphragm.

acoustic radiator—In an electroacoustic transducer, the part that radiates the sound vibration. Speaker cones and headphone diaphragms are examples.

acoustic radiometer—An instrument for measuring sound intensity by determining the unidirectional steady-state pressure caused by the reflection or absorption of a sound wave at a boundary.

acoustic reactance—That part of acoustic impedance due to the effective mass of the medium—that is, to the inertia and elasticity of the medium through which the sound travels. It is the imaginary component of acoustic impedance and is expressed in acoustic ohms.

acoustic refraction—A bending of sound waves as they pass obliquely from one medium to another in which the velocity of sound is different.

acoustic reflectivity—The ratio of the rate of flow of sound energy reflected from the surface on the side of incidence to the incident rate of flow.

acoustic regeneration— *See* Acoustic Feedback.

acoustic resistance—The acoustic analog of electrical resistance; that is, it is the opposition to transmission of sound energy through the medium. In air, it is the dissipation of energy by friction between adjacent molecules of air.

acoustic resonance—An increase in sound intensity as reflected waves and direct waves that are in phase combine. Resonance may also be caused by the natural vibration of air columns or solid bodies at a particular sound frequency.

acoustic resonator—An enclosure that intensifies those audio frequencies at which the enclosed air is set into natural vibration.

acoustics—1. The science of production, transmission, reception, and effects of sound. 2. In a room or other listening area, those characteristics that control reflection and absorption of sound waves and thus the sound reception in it.

acoustic system—Arrangement of components in devices designed to reproduce audio frequencies in a specified manner.

acoustic treatment—Use of certain sound-absorbing materials to control the amount of reverberation in a room, hall, or other enclosure.

acoustic wave—A traveling vibration by which sound energy is transmitted in air, in water, or in the earth. The characteristics of these waves may be described in terms of change of pressure, particle displacement, or density.

acoustic wave filter—A device designed to either attenuate or pass sound waves of different frequencies.

acoustoelectric effect—Generation of an electric current in a crystal by a traveling longitudinal sound wave.

ac transducer—A transducer that, for proper operation, must be excited with alternating currents only. Also, a device, the output of which appears in the form of alternating current.

ac voltage— *See* Alternating Voltage.

adapter—1. A fitting designed to change the terminal arrangement of a jack, plug, socket, or other receptacle, so that other than the original electrical connections are possible. 2. An intermediate device that permits attachment of special accessories or provides special means for mounting.

advance ball—In disc recording, a rounded support (often sapphire) that is attached to the cutting head and rides on the surface of the recording medium. Its purpose is to maintain a uniform mean depth of cut and to correct for small irregularities on the surface of the disc.

aeolight—A glow lamp that employs a cold cathode and a mixture of inert gases and in which the intensity of illumination varies with the applied signal voltage. This lamp is used to produce a modulated light for motion-picture sound recordings.

AES—Abbreviation for Audio Engineering Society.

af— *See* Audio Frequency.

afc— *See* Automatic Frequency Control.

agc— *See* Automatic Gain Control.

air column—The air space within a horn or acoustic chamber.

air suspension—A speaker design feature whereby part or all of the elasticity of the rim support of the speaker diaphragm is replaced by the elastic properties of an enclosed volume of air in the speaker enclosure.

alkaline cell—Also called alkaline-manganese cell. A primary cell similar to the zinc-carbon cell except that a potassium hydroxide electrolyte is used. Cell capacity is about 50% to 100% greater than with a zinc-carbon cell. Nominal cell voltage is 1.5 volts.

Allen screw—A screw having a hexagonal hole or socket in its head. Often used as a setscrew.

alligator clip—A spring-loaded metal clip with long, narrow, meshing jaws, used for making temporary electrical connections.

alnico—An alloy consisting mainly of ALuminum, NIckel, and CObalt plus iron. Capable of very high flux density and magnetic retentivity. Used in permanent magnets for speakers, magnetrons, etc.

alpha—Emitter-to-collector current gain of a transistor connected as a common-base amplifier. For a junction transistor, alpha is less than unity, or 1.

alternating current—Abbreviated ac. A flow of electricity that reaches maximum in one direction, decreases to zero, then reverses and reaches maximum in the opposite direction. The cycle is repeated continuously. The number of such cycles per second is the frequency, expressed in hertz.

alternating current/direct current—Abbreviated ac/dc. A term applied to electronic equipment that is capable of being operated from either an alternating-current or direct-current primary power source.

alternating-current erase head—An erasing head, used in magnetic recording to produce the magnetic field necessary for erasing. Alternating current is applied to the head, which then subjects the recording medium to a number of cycles of magnetic field of gradually decreasing magnitude.

alternating voltage—Also called ac voltage. Voltage that is continually varying in value and reversing in direction at regular intervals.

alternation—One-half of a cycle—either when an alternating current goes positive and returns to zero, or when it goes negative and returns to zero. Two alternations make one cycle.

alternator—A device for converting mechanical energy into electrical energy in the form of alternating current.

a-m— *See* Amplitude Modulation.

ambient noise—Acoustic noise in a room or other location. Literally, noise that surrounds on all sides.

American National Standards Institute—An independent, industry-wide association that establishes standards for the purpose of promoting consistency and interchangeability among the products of different manufacturers. Formerly the American Standards Association (ASA) and United States of America Standards Institute (USAS) Abbreviated ANSI.

am/fm receiver—A device capable of receiving and converting either amplitude- or frequency-modulated radio signals into audio-frequency signals.

am/fm tuner—A device capable of receiving and converting either amplitude- or frequency-modulated radio signals into low-level audio-frequency signals.

ammeter—An instrument for measuring either direct or alternating electrical current (depending on its construction). Its scale may be calibrated in amperes, milliamperes, microamperes, or kiloamperes.

amp—Abbreviation for ampere.

ampere—A unit of electrical current or rate of flow of electrons. One volt across 1 ohm of resistance causes a current of 1 ampere. A flow of 1 coulomb per second equals 1 ampere. The ampere is also defined as the constant current which, if maintained in two straight parallel conductors of infinite length, of negligible circular sections, and placed 1 meter apart in a vacuum will produce between these conductors a force equal to 2×10^{-7} newton per meter of length.

ampere-hour—A current of one ampere flowing for one hour. The unit usually used to measure the total amount of energy delivered by a battery.

ampere-turns—The number of turns of wire in a coil, multiplied by the current in amperes through the coil.

amplification—In an electrical or electronic device, the ratio of the output to the input. Amplification is often expressed in decibels.

amplification factor (μ)—In a vacuum tube, the ratio of a small change in plate voltage to the small change in grid voltage required to produce the same change in plate current (all other electrode voltages and currents being held constant).

amplifier—A device obtaining power from some source other than the input signal and which produces an output that is an enlarged reproduction of the input. The amplifying element is commonly an electron tube, a transistor, a magnetic circuit, or similar device.

amplitude distortion—In an amplifier or similar device, a type of distortion that is characterized by an output amplitude that is not a linear function of the input amplitude.

amplitude frequency response—The variation of gain, loss, amplification, or attenuation of a device or system as a function of frequency. Usually measured in the region where the transfer characteristic is essentially linear.

amplitude modulation—A system of modulation whereby the amplitude of the carrier wave is varied in step with amplitude changes of the modulating signal.

a-m suppression—The characteristic of an fm receiver that allows it to reject amplitude variations in the fm signal, thus contributing to quiet reception under adverse broadcasting conditions.

a-m tuner—A device capable of converting amplitude-modulated signals into low-level audio frequencies.

anechoic enclosure—A term originally applied to audio-frequency low-reflection enclosures. Now applied generally to low-reflection enclosures.

anechoic room—An enclosure in which reflection of sound energy is negligible. Used for measurement of microphone and speaker characteristics.

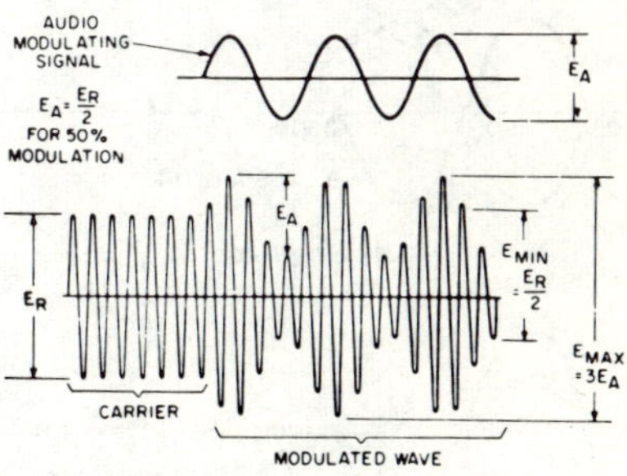

Amplitude modulation.

angle of lag (or lead)—The angular phase difference between one sinusoidal function and another having the same frequency. The function that passes through each amplitude variation at a later time than the other function is said to be lagging; the other function is said to be leading.

angstrom unit—A unit of measurement of the wavelength of light and other radiation. One angstrom is equal to one ten-thousandth of a micron or one hundred-millionth of a centimeter. The visible spectrum extends from about 4000 to 8000 angstrom units.

angular frequency—Frequency expressed in radians per second. One hertz equals 2π radians per second.

angular velocity—The rate at which an angle changes. Expressed in radians per second, the angular velocity of a periodic quantity is the frequency multiplied by 2π. If the periodic quantity results from uniform rotation of a vector, the angular velocity is the number of radians per second passed over by the rotating vector. Generally designated by the Greek letter omega (ω).

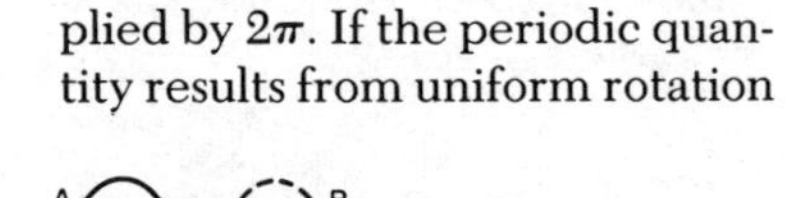

Angle of lag.

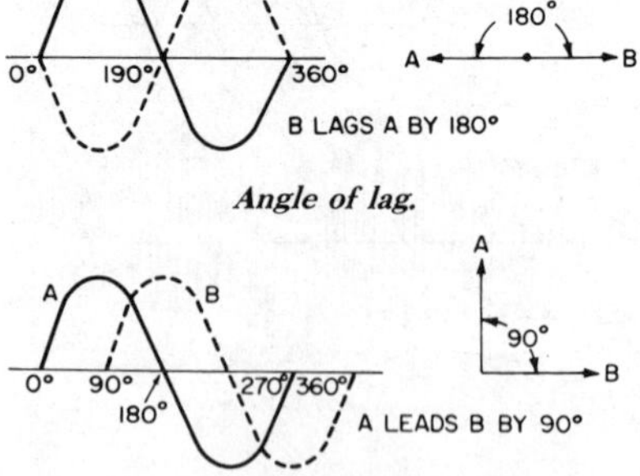

Angle of lead.

anode—The positive electrode, such as the plate of a vacuum tube.

anode current—The electron flow in the element designated as the anode. Usually signifies plate current of a vacuum tube.

anode voltage—The potential difference existing between anode and cathode.

antinodes—Also called loops. The points of maximum displacement in a series of standing waves. Two similar and equal wave trains traveling at the same velocity in opposite directions along a straight line result in alternate antinodes and nodes along the line. Antinodes are separated from their adjacent nodes by half a wavelength of the wave motion.

antinoise microphone—A microphone that discriminates against acoustic noise. A lip or throat microphone is an example.

antiresonant circuit—A parallel-resonant circuit offering maximum impedance to the series passage of the resonant frequency.

antiresonant frequency—The frequency at which the impedance of a parallel resonant system is maximum.

antiskating—A feature of some phonograph tone arms that counteracts the tendency of the arm to drift or "skate" toward the center of the record.

antistatic agents—Methods used to minimize static electricity in plastic materials. Antistatic agents are of two basic types: Metallic devices that come into contact with the charged material to conduct the charge to earth and chemical additives to the plastic. The first method can give complete neutralization while contact is maintained but does not prevent the material from becoming recharged if contact is broken. The second method gives more or less permanent protection if the additive is mixed with the plastic during manufacture.

aperiodic—Having no fixed resonant frequency or repetitive characteristics or no tendency to vibrate. A circuit that will not resonate within its tuning range is often called aperiodic.

aperiodic damping—Also called overdamping. The condition of a system having so much damping that when the system is subjected to a disturbance, either constant or instantaneous, the system comes to a position of rest, but does not pass through that position. An aperiodically damped system is not strictly an oscillating system, but it has such properties that it would become one if the damping were sufficiently reduced.

aperiodic function—A function having no repetitive characteristics.

apparent power—The product of voltage and current of a single-phase circuit in which the two reach their peaks at different times.

apparent source— *See* Effective Acoustic Center.

armature—The moving element in an electromechanical device such as the rotating part of a generator or motor, the movable part of a relay, or the spring-mounted, iron portion of a bell or buzzer.

armature gap—The space between the armature and the pole face.

artificial ear—A microphone-equipped device for measuring the sound pressures developed by an earphone. To the earphone it presents an acoustic impedance similar to the impedance presented by the human ear.

artificial voice—A device for simulating the natural properties of human speech. It consists of a small speaker mounted in an enclosure designed to duplicate as closely as possible the acoustical constants of the human head. It is useful for testing and calibrating close-talking microphones.

ASI— *See* American National Standards Institute.

ASRA—Abbreviation for automatic stereophonic recording amplifier. An instrument developed by Columbia Broadcasting System for stereo recording. Compression of the vertical component of the stereo recording signal is automatically increased or decreased as required by the recording conditions.

attachment cord— *See* Patch Cord.

attack—The length of time it takes for a tone in an organ to reach full intensity after a key is pressed. On most organs this effect is adjustable by either a switch or a potentiometer.

attenuation—1. The decrease in amplitude of a signal during its transmission from one point to another. It may be expressed as a numerical ratio, or in decibels. 2. *See* Insertion Loss.

attenuation network—1. A network providing relatively slight phase shift and substantially constant attenuation over a range of frequencies. 2. The arrangement of circuit elements, usually impedance elements, inserted in circuitry to introduce a known loss or to reduce the impedance level without reflections.

attenuator—1. A resistive network that provides reduction of the amplitude of an electrical signal without introducing appreciable phase or frequency distortion. 2. A distributed network that absorbs part of a signal and transmits the remainder with a minimum of distortion or delay.

atto—Prefix meaning 10^{-18}.

audibility—1. The ability to be heard, usually construed as being heard by the human ear. 2. The ratio of the level of a specific sound to the level of a sound that can barely be heard. Usually expressed in decibels.

audible—Capable of being heard, in most contexts, by the average human ear.

audible tones—Sounds that the normal human ear can detect. The frequency range normally accepted is 30 to 15,000 Hz.

audio—Pertaining to frequencies corresponding to a normally audible sound wave. These frequencies range roughly from 15 hertz to 20,000 hertz.

audio amplifier— *See* Audio Frequency Amplifier.

audio band—The range of audio frequencies passed by an amplifier, receiver, transmitter, etc.

audio component—That portion of any wave or signal that contains

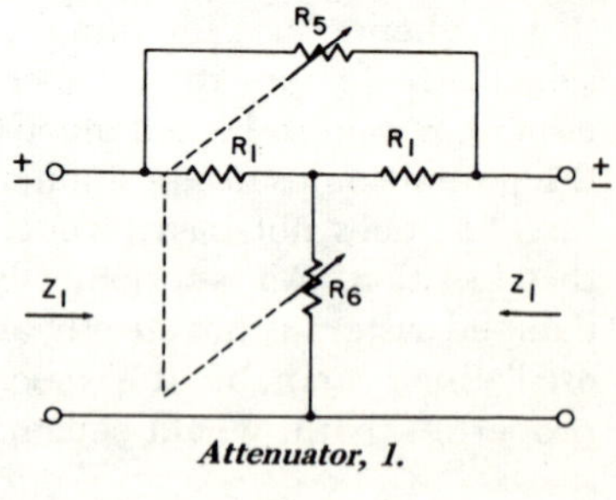

Attenuator, 1.

frequencies in the audible range (between 15 hertz and 20,000 hertz).

audio frequency—Abbreviated af. Any frequency corresponding to a normally audible sound wave. The audio frequency range is generally considered to extend from 15 hertz to 20,000 hertz.

audio-frequency amplification—An increase in voltage, current, or power of a signal at an audio frequency.

audio-frequency amplifier—Also called audio amplifier. A device containing amplifying elements (tubes, transistors, or others) designed to amplify signals within a frequency range of about 15 hertz to 20,000 hertz.

audio-frequency choke—An inductance used to offer impedance at audio frequencies.

audio-frequency noise—In the audio-frequency range, an electrical disturbance introduced from a source extraneous to the signal.

audio-frequency oscillator—An oscillator circuit using an electron tube, transistor, or other nonrotating device capable of producing audio signals.

audio-frequency transformer—Also called audio transformer. An iron-core transformer for use with audio-frequency currents to transfer signals from one circuit to another. Used for impedance matching or to permit maximum transfer of power.

audiogram—Also called threshold audiogram. A graph showing hearing loss, percent of hearing loss, or percent of hearing as a function of frequency.

audio-level meter—An instrument that measures audio-frequency power with reference to a predetermined level. Usually calibrated in decibels.

audiometer—An electronic instrument for measuring a listener's hearing acuity. In simple units, the listener is provided (usually through earphones) with an audio signal (commonly a pure tone) of known intensity and frequency. More complex instruments provide a variety of signals (pure tones, white noise, and speech) through a variety of output transducers such as earphones, bone vibrators, or loudspeakers.

audio oscillator— *See* Audio-Frequency Oscillator.

audiophile—A person who is interested in good musical reproduction for his own personal listening and who uses the latest audio equipment and techniques.

audio signal—An electrical signal of a frequency within the audio range.

audio spectrum—The continuous range of audio frequencies extending from the lowest to the highest (from about 15 hertz to 20,000 hertz). The illustration

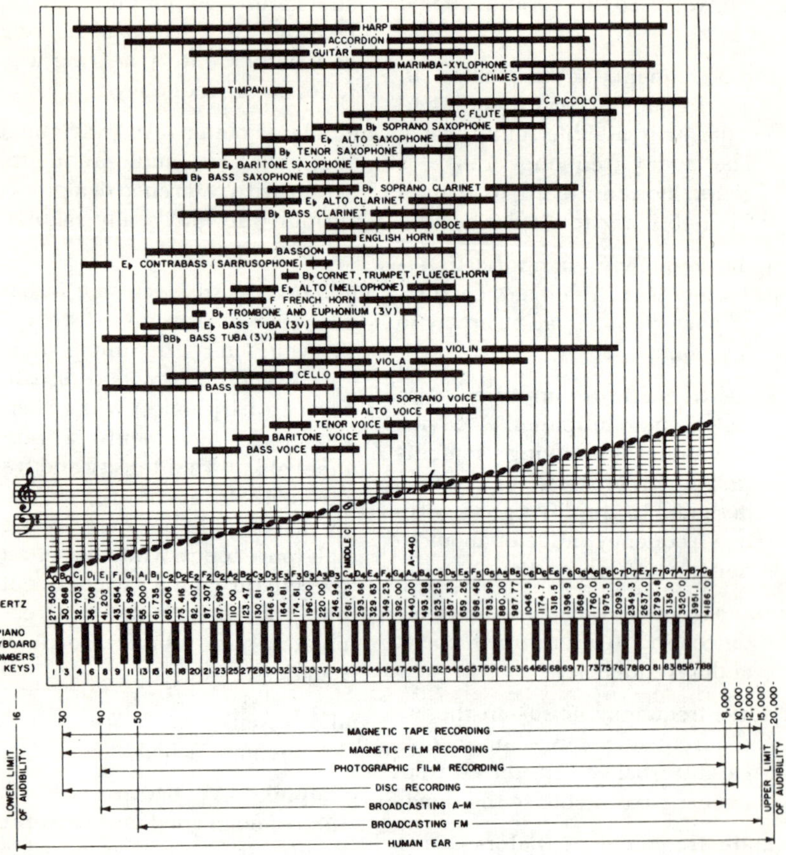

Audio spectrum.

shows the range of various musical instruments and recording and broadcast systems.

audio taper—Term used to describe volume controls and potentiometers. Semilogarithmic change of resistance with control rotation. Used on tone controls to compensate for the lower sensitivity of the human ear when listening to low-volume sounds.

audio transformer— *See* Audio-Frequency Transformer.

audiovisual—Involving both sight and sound.

audiovisual system—A system of communication that simultaneously transmits pictorial and audio signals.

aural—Pertaining to the ear or to the sense of hearing. (*See also* Audio.)

automatic bass compensation—A circuit used in a receiver or audio amplifier to make bass notes sound more natural at low volume settings. The circuit, which usually consists of resistors and capacitors connected to taps on a volume control, automatically compensates for the poor response of the human ear to weak sounds.

automatic frequency control—Abbreviated afc. A circuit that automatically maintains the frequency of an oscillator within specified limits.

automatic gain control—Abbreviated agc. A circuit that automatically maintains the output volume of a receiver constant, regardless of variations in the signal strength applied to a receiver.

automatic noise limiter—A circuit that automatically clips all noise peaks above a certain desired signal level. This circuit prevents strong atmospheric or man-made interference from being troublesome.

automatic record changer—An electrically operated mechanism that automatically feeds, plays, and rejects a number of records in a preset sequence.

automatic shutoff—Provision for automatically stopping a tape recorder at end of tape or in case of tape breakage. Also, a switching arrangement that shuts off a record player at the end of the last record of a stack.

automatic tuning—An electrical, mechanical, or electrical/mechanical system that automatically tunes a circuit to a predetermined frequency when activated by a button or other type of control.

automatic volume compression—*See* Volume Compression.

automatic volume control (avc)—A device that automatically maintains the output of a radio receiver or amplifier substantially constant within relatively narrow limits while the input level varies over a wide range.

automatic volume expansion—Also called volume expansion. An audio-frequency circuit that automatically increases the volume range by making loud portions louder and weak ones weaker. This is done to make radio reception sound more like the actual program because the volume range of programs is generally compressed at the point of broadcast.

autotransformer—1. A transformer with a single winding (electrically) in which the entire winding acts as the primary winding, and only part of the winding acts as the secondary (step-down); or part of the winding acts as the primary and the whole winding acts as the secondary (step-up). 2. A voltage, current, or impedance-transforming device in which parts of one winding are common to both the primary and the secondary circuits.

avalanche breakdown—In a semiconductor diode, a nondestructive breakdown caused by the cumulative multiplication of carriers through field-induced impact ionization.

avalanche conduction—A form of conduction in a semiconductor in which charged-particle collisions create additional hole-electron pairs.

avalanche diode—Also called breakdown diode. A silicon diode that has a high ratio of reverse-to-forward resistance until avalanche breakdown occurs. After breakdown, the voltage drop across the diode is essentially constant and is independent of the current. Used for voltage regulating and voltage limiting.

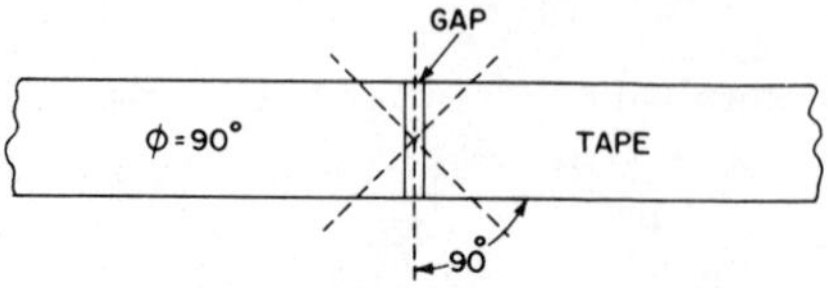

Azimuth alignment (magnetic head).

Originally called zener diode, before it was found that the Zener effect had no significant role in the operation of diodes of this type.

avc— *See* Automatic Volume Control.

average value—The value obtained by dividing the sum of a number of quantities by the number of quantities. The average value of a sine wave is 0.637 times the peak value.

azimuth alignment—In a tape recorder, adjustment of the angle the record or playback gap of the head makes with the direction of tape travel. This angle should be 90 degrees in a properly aligned head.

B

back current—Also called reverse current. The current that flows when reverse bias is applied to a semiconductor junction.

back diode—A tunnel diode that is usually chosen for its reverse-conduction characteristics.

background noise—Any unwanted sound that intrudes upon program material; for example, the sounds produced as a result of surface imperfections of a disc record.

backloaded horn—A speaker enclosure arranged so the sound from the front of the cone feeds directly into the room, while the sound from the rear feeds into the room via a folded horn.

baffle—In acoustics, a shielding structure or partition used to increase the effective length of the external transmission path between two points (for example, between the front and back of an electroacoustic transducer). A baffle is often used with a speaker to increase the acoustic loading of the speaker diaphragm.

balance—1. The effect of blending the volume of various sounds coming over different microphones in order to present them in correct proportion. 2. The maintenance of equal average volume from both speaker systems of a stereo system.

balance control—On a stereo amplifier, a differential gain control used to vary the volume of one speaker system relative to the other without affecting the overall volume level.

balanced amplifier—An amplifier circuit with two identical signal branches, connected to operate in opposite phase and with their input and output connections each balanced to ground; for example, a push-pull amplifier.

banana jack—A jack that accepts a banana plug. Generally designed for panel mounting.

banana plug—A plug with a banana-shaped spring-metal tip and with elongated spring, to provide a low-resistance compression fit.

band-elimination filter— Also called band-stop filter. A wave filter with a single attenuation band, neither of the cutoff frequencies being zero or infinite. frequencies. This is often done by attenuating the higher frequencies, either by selective negative feedback or by a filter network.

The filter passes frequencies on either side of this band.

bandpass— The difference between the upper and lower frequency limits at which a desired signal level is passed (usually all levels from maximum down to half-power of maximum).

bandpass amplifier circuit— A stage designed to uniformly amplify signals over a certain continuous range of frequencies only.

bandpass filter—A wave filter with a single transmission band, neither of the cutoff frequencies being zero or infinite. The filter attenuates frequencies on either side of this band.

bandpass response—Also called flat-top response. The response characteristic in which a definite band of frequencies is transmitted uniformly.

band-reject filter—A filter that does not pass a certain band of frequencies but passes higher and lower frequencies. Sometimes called a notch filter.

band-stop filter— *See* Band-Elimination Filter.

bandwidth—The range within the limits of a specific band. The width of a bandpass filter is generally taken as the limits between which its attenuation is not more than 3.0 decibels greater than its average attenuation throughout its passband.

barium titanate—A ceramic that has piezoelectric properties and is more resistant to high temperatures and high humidity than Rochelle salt crystals. Used in microphones, phonograph pickups, and sonar transducers.

base—The region between the emitter and collector of a transistor that receives minority carriers injected from the emitter.

base film—The plastic substrate that supports the coating of magnetic recording tape. The base film of most instrumentation and computer tapes is made of polyester. Other bases used for less critical applications are cellulose acetate and polyvinyl chloride.

base voltage—The voltage between the base terminal of a transistor and the reference point.

bass—Sounds in the low audio frequency range. On the standard piano keyboard, all notes below middle C (261.63 hertz).

bass boost—A deliberate adjustment of the amplitude/frequency response of a system or component to accentuate the lower audio frequencies.

bass-boost circuit—A circuit that emphasizes the lower or bass frequencies relative to the higher frequencies. This is often done by attenuating the higher frequencies, either by selective negative feedback or by a filter network.

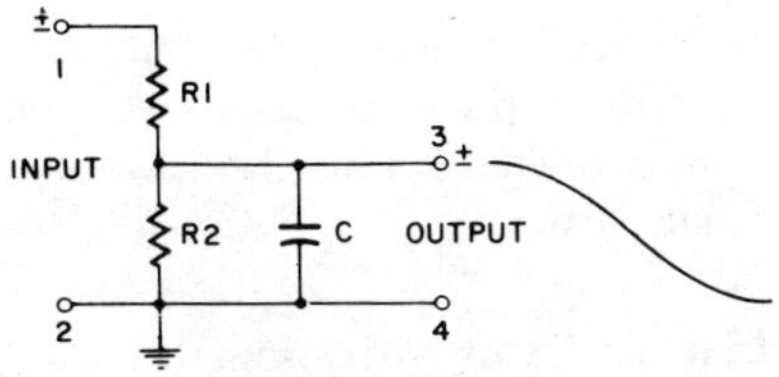

Bass-boost circuit.

bass compensation—Emphasizing the low-frequency response of an audio amplifier at low volume levels to compensate for the lowered sensitivity of the human ear to weak low frequencies.

bass control—A manual tone control that has the effect of changing the level of bass frequencies reproduced by an audio amplifier.

bass-reflex enclosure—A type of speaker enclosure in which the rear wave from the speaker emerges through an auxiliary opening or port of critical dimensions to reinforce the bass tones. (*See also* Vented Baffle.)

bass response—The extent to which a speaker or audio-frequency amplifier handles low audio frequencies. 2. The ability of any device to pick up or reproduce low audio frequencies.

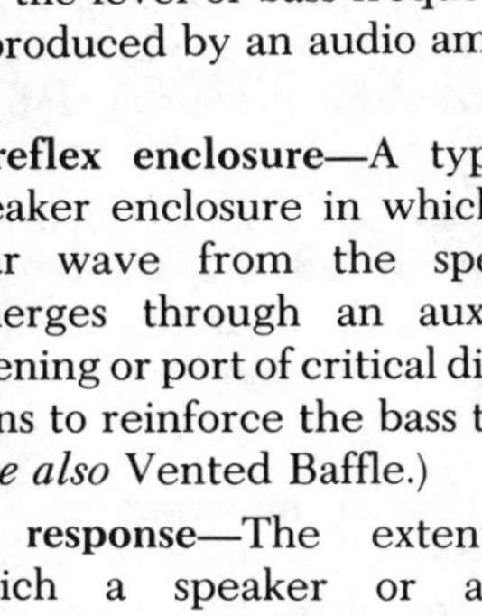

Bass reflex enclosure.

battery—A dc voltage source consisting of two or more cells that convert chemical, nuclear, solar, or thermal energy into electrical energy.

battery capacity—The amount of energy obtainable from a storage battery, usually expressed in ampere-hours.

B channel—One of two stereo channels, usually the right, together with microphone, speakers, or other equipment associated with this channel.

beat frequency—The frequency of the alternate reinforcing and cancelling action that occurs between two waves of slightly different frequency. The beat frequency is equal to the difference between the two frequencies.

beat note—The difference frequency produced when two sinusoidal waves of different frequencies are applied to a nonlinear device.

bel—The fundamental unit in a logarithmic scale for expressing the ratio of two amounts of power. The number of bels is equal to the $\log_{10} P_1/P_2$, where P_1 is the power level being considered and P_2 is an arbitrary reference level. The decibel, equal to 1/10 bel, is a more commonly used unit.

beta—Symbolized by the Greek letter *beta* (β). Also called current transfer ratio. The current gain of a transistor connected as a grounded-emitter amplifier; it is the ratio of a small change in collector current to the corresponding change in base current, with the collector voltage constant.

B-H curve—Curve plotted on a graph to show successive states during magnetization of a ferromagnetic material. A normal magnetization curve is a portion of a symmetrical hysteresis loop. A virgin magnetization curve shows what happens the first time the material is magnetized.

B-H meter—An instrument for measuring the intrinsic hysteresis loop of a sample of magnetic material.

bias— The electrical, mechanical, or magnetic force applied to a relay, vacuum tube, or other device for the purpose of establishing an electrical or mechanical reference level for the operation of the device.

bias cell—A dry cell used in a circuit to provide bias voltage.

bias current—The steady-state current through the base-emitter junction of a transistor. Its value is selected to set the operating point of the transistor.

bias distortion—Distortion resulting from operation on a nonlinear portion of the characteristic curve of a vacuum tube or other device, because of improper biasing.

bias-induced noise—In tape recording, the difference between bulk-erased and zero-modulation noise.

bias oscillator—An oscillator used in magnetic recorders to generate an ac signal in the range of 40 kHz to 80 kHz for the purpose of magnetic biasing to obtain a linear recording characteristic. Usually the bias oscillator also serves as the erase oscillator.

bias resistor—A resistance used in a vacuum-tube or transistor circuit to provide bias voltage or current for proper operation.

bias, tape—A high-frequency current applied to a tape recording circuit to minimize distortion during recording.

bidirectional microphone—A microphone in which the response predominates for sound sources having polar coordinates of 0° and 180°

bimorph cell—Two crystal elements (usually Rochelle salt) in rigid combination, arranged to act as a mechanical transformer in headphones, microphones, pickups, and speakers.

binaural—Two-eared; applied to a kind of recording that is reproduced by separate speakers or headphones to create a sense of depth.

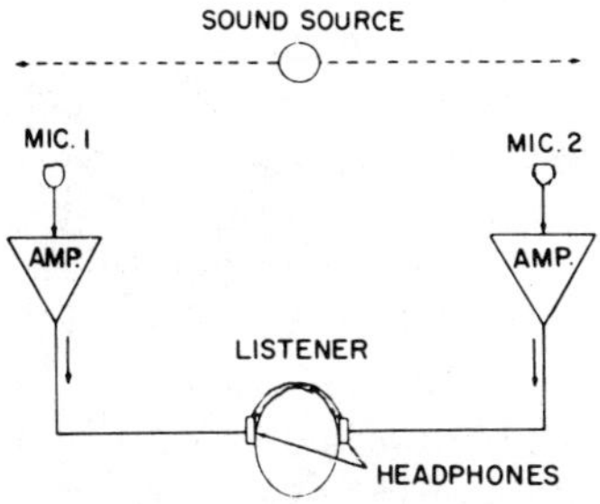

Binaural-sound reproducing system.

binaural disc—A stereo record with two separate signals recorded in its grooves. Stereophonic sound is obtained by feeding each signal into its own speaker or headphone.

binaural effect—The effect that makes it possible for a person to determine the spatial position of a sound by differences of intensity and arrival time at right and left ears.

binaural recorder—A recording system that employs two separate recording channels, each with its own microphone, amplifier, recording and playback heads, and earphones. For true binaural effect, earphones are required; although the term is sometimes applied to a system using right and left speakers with separate signals applied to each speaker, the term is not correctly used, inasmuch as each ear hears the signal from both speakers.

binaural recording—Recording with two microphones on separate tape tracks, to be played back through separate amplifiers and binaural headphones.

binaural sound—Sound recorded and reproduced by equipment arranged in pairs of right and left so that the spatial effect of the original sound is produced.

bipolar device—A semiconductor device in which there are both majority and minority carriers. (This is the case in all npn and pnp transistors.)

bipolar magnetic driving unit—A headphone or speaker unit having two magnetic poles acting directly on a flexible iron diaphragm.

bipolar transistor—A transistor that uses both negative and positive charge carriers.

blank groove— *See* Unmodulated Groove.

blank tape— *See* Raw Tape.

blasting—Overloading of an amplifier or speaker, resulting in severe distortion of loud sounds.

blocked impedance—The input impedance of a transducer when its output is connected to a load of infinite impedance.

blocked resistance—Resistance of an audio-frequency transducer when its moving elements are restrained so they cannot move; it represents the resistance due only to electrical loss.

boffle—A speaker enclosure, developed by H.A. Hartley, containing a group of stretched, resilient, sound-absorbing screens.

bonded pickup— *See* Bonded Transducer.

bonded strain gauge—A pressure transducer that uses a pressure-sensing system consisting of strain-gauge elements firmly bonded to a pressure-responsive member. Thermal stability and insensitivity to shock and vibration are improved by this bonded construction.

bonded transducer—Also called bonded pickup. A transducer that employs the bonded strain-gauge principle of transduction.

bone conduction—The process by which sound is conducted to the inner ear through the cranial bones.

boom—A mechanical support for a microphone, used in a television studio to suspend the microphone within range of the speakers' voices, but out of camera range.

breakdown voltage—That voltage at which an insulator or dielectric ruptures, or at which ionization and conduction take place in a gas or a vapor.

break elongation—The relative elongation of a recording tape or base film at the instant it breaks after having been stretched at a given rate.

brilliance—The degree to which the higher audio frequencies resemble the original when reproduced by a receiver or public-address amplifier.

brilliance control—A potentiometer used in a three-way speaker system to adjust the output level of the tweeter for proper volume relative to the lower audio frequencies produced by the speaker system.

broad-band amplifier—An amplifier that has an essentially flat response over a wide frequency range.

broad-band electrical noise—Also called random noise. A noise signal that contains a wide range of frequencies and has a randomly varying instantaneous amplitude.

bulk degausser— *See* Bulk Eraser.

bulk-erased noise—The noise arising when a bulk-erased tape is reproduced with the erase and record heads completely de-energized. Ideally, this noise is governed by the number of magnetic particles that pass by the head in unit time.

bulk eraser—Also called bulk degausser or degausser. Equipment for erasing a complete roll of tape in one operation. The eraser subjects the tape roll to a gradually decreasing 60-hertz ac field, either by withdrawing the roll from the field or by reducing the input to the eraser.

Burwen system—A noise reduction system that compresses dynamic range during recording and expands it during playback.

C

C—Symbol for capacitor, capacitance, coulomb, centigrade or Celsius, and transistor collector.

cactus needle—A phonograph needle made from the thorn of a cactus plant.

cadmium cell—A standard cell used as a voltage reference; at 20° C its voltage is 1.0186 volts.

calibrate—To ascertain, by measurement or by comparison with a standard, any variations in the readings of another instrument, or to correct the readings.

calibration—1. The process of comparing an instrument or device with a standard to determine its accuracy or to devise a corrected scale. 2. The taking of measurements of various parts of electronic equipment to determine the performance level of the equipment and whether it conforms to technical order specifications.

capacitance—Abbreviated C. Also called capacity. In a capacitor or a system of conductors and dielectrics, that property that permits the storage of electrically separated charges when potential differences exist between the conductors. The capacitance of a capacitor is defined as the ratio between the electric charge that has been transferred from one electrode to the other and the resultant difference in potential between the electrodes. The value of this ratio depends on the magnitude of the transferred charge.

$$C\ (\text{farad}) = \frac{Q\ (\text{coulomb})}{V\ (\text{volt})}$$

capacitive coupling—Also called electrostatic coupling. The association of two or more circuits with one another by means of a connective capacitor.

capacitive reactance—Symbolized by X_C. The impedance of a capacitor to ac or pulsating dc. Measured in ohms and equal to

$$\frac{1}{2\pi f C}$$

where,
f is the frequency in hertz,
C is the capacitance in farads.

capacitive speaker— *See* Electrostatic Speaker.

capacitor—A device consisting essentially of two conducting surfaces separated by an insulating material or dielectric such as air, paper, mica, glass, plastic film, or oil.

capacitor microphone— *See* Electrostatic Microphone.

capacitor pickup—A phonograph pickup that depends for its operation on the variation of its electrical capacitance with stylus movement.

capacitor speaker— *See* Electrostatic Speaker.

capstan—The driven spindle or shaft in a tape recorder—sometimes the motor shaft itself—that rotates against the tape, pulling it through the machine at constant speed during recording and playback modes. The rotational speed and diameter of the capstan thus determine the tape speed.

capstan idler— *See* Pressure Roller.

carbon microphone—A microphone that depends on the variation in resistance of carbon contacts for its operating effects.

cardioid diagram—Heart-shaped polar diagram showing the response or radiation characteristic of certain directional antennas, or the response characteristic of certain types of microphones.

cardioid microphone—A microphone with a heart-shaped response pattern that gives nearly uniform response over about 180° at the front of the microphone and minimum response at the rear.

carillon—A bell tower designed to play from a keyboard. In an organ, bell-like tones may be achieved by metal tubes struck with felt hammers, or they may be achieved solely by electronic synthesis.

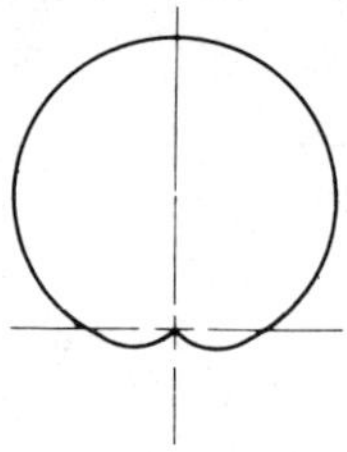

Cardioid diagram.

carrier—A wave of constant amplitude, frequency, and phase that can be modulated by changing amplitude, frequency, or phase in step with a modulating signal.

carrier frequency—1. The frequency (in hertz) of the wave modulated by the intelligence wave. 2. The reciprocal of the period of a carrier. 3. The frequency of the unmodulated fundamental output from a radio transmitter.

carrier-frequency stereo disc—A stereo disc with two laterally cut channels having the following characteristics. One channel is cut in the usual manner. The second channel is used to frequency-modulate a supersonic carrier frequency. The playback cartridge delivers the signal for one channel plus the carrier frequency containing the signal from the other channel. The latter signal is obtained by demodulation.

Cassette.

carrier system—A means of transmitting a number of channels over a single path by modulating each channel upon a different carrier frequency and demodulating at the receiving point to restore the signals to their original form.

cartridge—A removable phonograph pickup.

cascade—An arrangement of two or more similar circuits or amplifying stages in which the output of one circuit provides the input of the next. Also called tandem.

cascade amplifier—A multiple-stage amplifier in which the output of each stage is connected to the input of the next stage.

cassette—A small plastic container that encloses two reels (tape supply and takeup) and that provides openings for heads and tape-drive mechanism. The entire unit is easily and quickly plugged into a cassette player. Standard tape speed for cassettes is 1 ⅞ inches per second.

cassette player—A machine designed for accepting and playing prerecorded tape cassettes. Does not contain circuitry for recording.

cassette recorder—A machine for recording and playing tape cassettes.

cassette track layout—Arrangement of recorded tracks on a cassette tape. (*See* illustration.)

cathode—1. In an electron tube, the electrode that is the primary source of electrons. 2. General name for any negative electrode. 3. When a semiconductor diode is biased in the forward direction, that terminal of the diode that is negative with respect to the other terminal.

cathode bias—A method of biasing a vacuum tube by placing the

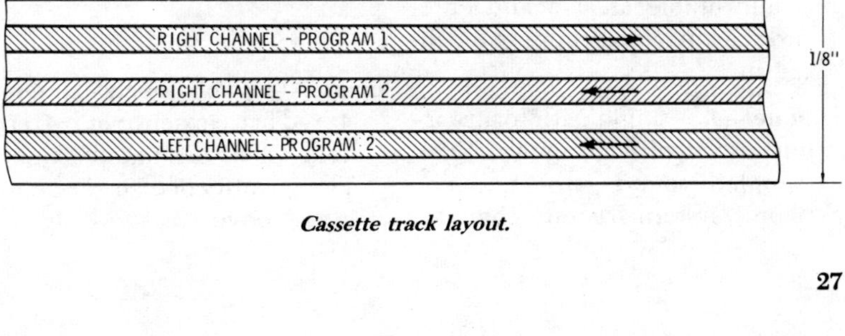

Cassette track layout.

biasing resistor in the common cathode-return circuit, thereby making the cathode more positive—rather than the grid more negative—with respect to ground.

cathode follower—Also called grounded plate amplifier. A vacuum-tube circuit in which the input signal is applied to the control grid and the output is taken from the cathode. The equivalent circuit using a transistor is called an emitter follower.

CD-4—A discrete 4-channel disc developed for quadraphonic sound by Japan Victor Company (JVC) and RCA.

cellulose-nitrate disc— *See* Lacquer Disc.

ceramic microphone—A microphone with a ceramic element.

ceramic pickup—A phonograph pickup with a ceramic cartridge.

certified magnetic tape—Magnetic tape that has been tested and is certified to be free of defects over its entire recording surface.

changer—A device that plays several phonograph records in sequence automatically. It consists of a turntable, an arm, and a record stacking and dropping mechanism.

channel—l. A single path for transmitting electrical signals. 2. A complete sound path. A monophonic system has one channel; a stereophonic system has at least two full channels.

channel balance—Equal response on both right and left channels.

channel reversal—Shifting the outputs of a stereo system so the channel formerly heard from the left speaker now comes from the right and vice versa.

channel-reversing switch— A switch that reverses the connections of two speakers in a stereo system with respect to the channels, so that the channel heard previously from the right speaker is heard from the left and vice versa.

channel separation—In stereo, the electrical or acoustical difference between the right and left channels. Inadequate separation can lessen the stereo effect; excessive separation can exaggerate it beyond natural proportions.

characteristic curve—A waveform or graph showing how some value or property of a device varies as some other value or aspect of the device is varied. Often applied to transducers and recording systems. (*See also* RIAA curve.)

charge—1. The electrical energy stored in a capacitor or battery or held on an insulated object. 2. The quantity of electrical energy in (1) above.

charge carrier—A mobile hole or conduction electron in a semiconductor.

charge density—The charge per unit area on a surface or charge per unit volume in space.

charged particle—An ion; an elementary particle that carries a positive or negative electric charge.

chatter—The vibration of a cutting stylus in a direction other than that in which it is driven.

chip—1. Also called thread. In mechanical recording, the material removed from the recording medium by the recording stylus as it cuts the groove. 2. A single substrate on which all the active and passive elements of an electronic circuit have been fabricated, using the semiconductor technologies of diffusion, passivation, masking, photoresist, and epitaxial growth.

chord—A harmonious combination of tones sounded together through the use of one or more fingers on either or both hands.

chord organ—An organ with provision for playing a variety of chords, each produced by means of a single button or key.

Christmas-tree pattern— *See* Optical Pattern.

circuit—1. An electronic path between two or more points capable of providing a number of channels. 2. A number of conductors connected together for the purpose of carrying an electrical current. 3. The interconnection of a number of components in one or more closed paths to perform a desired electrical or electronic function.

clavier—Any keyboard, either hand or foot operated.

close-talking microphone—Also called noise-cancelling microphone. A microphone designed to be held close to the mouth of the operator.

coating—The magnetic layer, consisting of oxide particles held in a binder that is applied to the base film, used for magnetic recording.

coating thickness—The thickness of the magnetic coating applied to the base film. Modern tape coatings range in thickness from 170 to 650 microinches, with a preponderance of coatings being approximately 400 microinches thick. In general, thin coatings give good resolution but lose in output at long wavelengths; thick coatings give high output at long wavelengths but lose in resolution.

coaxial speaker—1. A single speaker comprising a high- and a low-frequency unit, plus an electrical crossover network. 2. A tweeter mounted on the axis of and inside the cone of a woofer.

coefficient of reflection—The square root of the ratio of the power reflected from a surface to the power incident on the same surface.

coercive force—Symbolized by H. The magnetizing force that must be applied to a magnetic material in a direction opposite to the residual induction to reduce the induction to zero.

coercivity—The property of a magnetic material measured by the coercive force corresponding to the saturation induction for the material.

collector—The collector of a transistor is an electrode through which a primary flow of carriers leaves the interelectrode region.

combination microphone—A microphone consisting of two or more similar or dissimilar microphones combined into one.

combination tones—Frequencies produced in a nonlinear device such as in an audio amplifier having appreciable harmonic distortion.

compander—Literally, compressor/expander. A combination consisting of a compressor at one point of a communication path to reduce the volume range of signals, followed by an expander at another point to improve the ratio of the signal to the interference entering the path between the compressor and expander.

companding—A noise reduction process that employs compression followed by expansion. Compression is applied before exposure to noise-generating conditions and expansion is applied after the exposure.

compatibility—The degree to which a specially designed system can be used in whole or in part with another system or part without unsatisfactory results. Examples: mono record player with a stereo pickup; quadraphonic recording playable on mono or stereo equipment; quadraphonic sound broadcast with existing fm/multiplex systems or in monophonic mode over existing a-m radio systems.

compensated amplifier—A broadband amplifier with a frequency range extended by the proper choice of circuit constants.

compensated volume control— *See* Loudness Control.

compensator—An electronic circuit for altering the frequency response of an amplifier system to achieve a specified result. This refers to record equalization or loudness correction.

complex tone—A sound wave produced by the combination of simple sinusoidal components of different frequencies. A sound sensation characterized by more than one pitch.

complex wave—A periodic wave made up of a combination of several frequencies or several sine waves superimposed on one another.

complex-wave generator—A device that generates a nonsinusoidal signal having a desired repetitive characteristic and waveform.

compliance—l. The reciprocal of stiffness—i.e., the ability to yield or flex. 2. The ease with which a phonograph stylus responds to an outside force. 3. The mechanical and acoustical equivalent of capacitance.

component—1. An essential functional part of a subsystem or equipment. It may be any self-contained element with a specific function, or it may consist of a combination of parts, assemblies, accessories, and attachments. 2. In vector analysis, one of the parts of a wave, voltage, or current considered separately.

compound horn—An electromagnetic horn of rectangular cross section. The four sides of the horn diverge in such a way that they coincide with or approach four planes, with the provision that the two opposite planes do not intersect the remaining planes.

compressed-air loudspeaker—A speaker that has an electrically actuated valve to modulate a stream of compressed air.

compression driver unit—A speaker driver unit that does not radiate directly from the vibrating surface. Instead, it requires acoustic loading from a horn that connects through a small throat to an air space adjacent to the diaphragm.

compressor—A transducer that for a given amplitude range of input voltages produces a smaller range of output voltages. In one important type of compressor, the envelope of speech signals is used to reduce their volume range.

concentric groove— *See* Locked Groove.

condenser microphone— *See* Electrostatic Microphone.

condenser speaker— *See* Electrostatic Speaker.

conductor—1. A bare or insulated wire, or a combination of wires not insulated from one another, suitable for carrying an electric current. 2. A body of conductive material so constructed that it will serve as a carrier of electric currrent. 3. A material (usually a metal) that conducts electricity through the transfer of orbital electrons.

cone—The diaphragm that sets the air in motion to create a sound wave in a direct-radiator speaker. Usually it is conical in shape.

conical horn—A horn the cross-sectional area of which increases as the square of the axial length.

connector—1. A coupling device that provides an electrical or mechanical junction between two cables, or between a cable and a chassis or enclosure. 2. A device

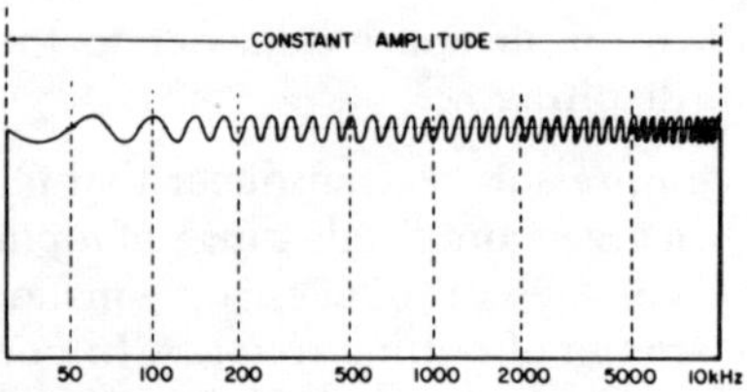

Constant-amplitude recording (characteristic of a crystal cutting head).

that provides rapid connection and disconnection of electrical cable and wire terminations.

consonance—Electrical or acoustical resonance between bodies or circuits not connected directly together.

constant-amplitude recording—A disc recording made with a recording characteristic that produces a groove displacement of constant amplitude when the signal input is of constant amplitude, regardless of the frequency of the input signal.

constant-k filter—A wave filter in the formula of which a constant term *k* appears. The impedance of the filter matches the line impedance at only one frequency.

constant-velocity recording—The disc-recording characteristic in which groove displacement is inversely proportional to signal frequency.

contact microphone—A microphone designed to pick up mechanical vibrations directly from the sound source and convert them into corresponding electrical currents or voltages.

continuous power—The power an amplifier is capable of delivering for at least 30 seconds with a sine-wave signal.

Cook system—An early stereo-disc recording technique in which

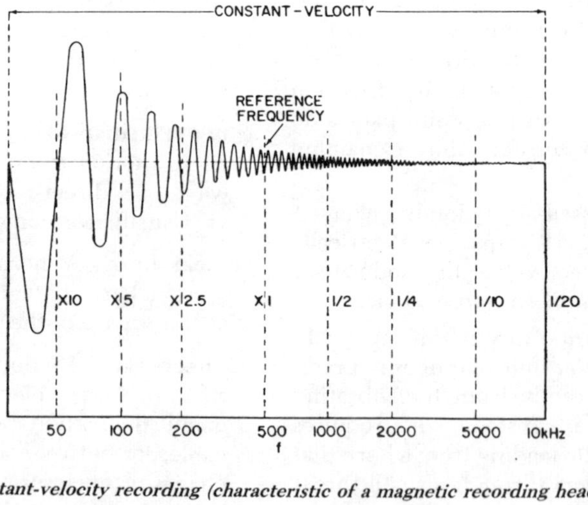

Constant-velocity recording (characteristic of a magnetic recording head).

the two channels were recorded simultaneously with two cutters on different portions (bands) of a record as concentric spirals. The playback equipment consisted of two pickups mounted side by side so that each played at the correct spot on its own band.

core—Magnetic material placed within a coil to intensify the magnetic field.

core loss—Also called iron loss. Loss of energy in a magnetic core as the result of eddy currents, which circulate through the core and dissipate energy in the form of heat.

corrective equalization— *See* Frequency-Response Equalization.

coulomb—The quantity of electricity that passes any point in an electric circuit in 1 second when the current is maintained constant at 1 ampere. The coulomb is the unit of electric charge in the mksa system. A charge of 1 coulomb contains 6.24×10^{18} electrons.

Coulomb's law—Also called law of electric charges or law of electrostatic attraction. The force of attraction or repulsion between two charges of electricity concentrated at two points in an isotropic medium is proportional to the product of their magnitudes and inversely proportional to the square of the distance between them. The force between unlike charges is an attraction and between like charges is a repulsion.

counter—A device driven by the tape-recorder mechanism that indicates the amount of tape used.

cps—Abbreviation for cycles per second. Now replaced by hertz (Hz).

critical damping—The value of damping that provides the most rapid transient response without overshoot. Occurs between underdamping and overdamping.

Crosby system—A compatible multiplex fm stereo broadcast technique in which the right and left signals are combined in phase (sum signal) and transmitted on the main carrier, and also combined out-of-phase (difference signal) and transmitted on the subcarrier. The two signals are combined (matrixed) in the receiving apparatus to restore the right and left channels.

cross coupling—Unwanted coupling between two different communication channels or their components.

crossover distortion—Distortion that occurs in a push-pull amplifier at the points of operation where the input signals cross over (go through) the zero reference points.

crossover frequency—As applied to electrical dividing networks, the frequency at which equal power is delivered to each of the adjacent frequency channels when all channels are terminated in the specified load.

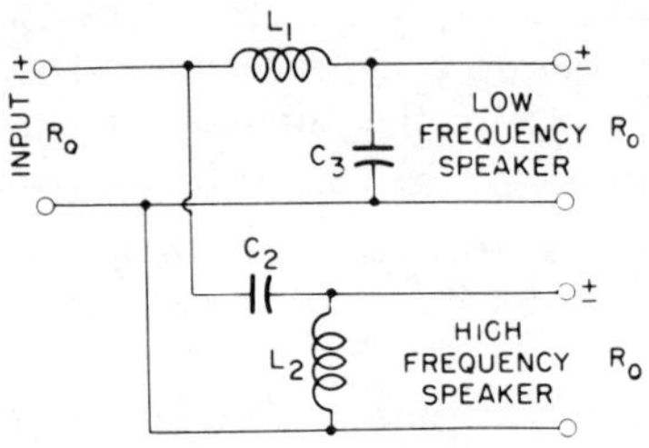

Crossover network.

crossover network—An electrical filter that separates the signal of an audio amplifier into two or more separate frequency bands for a multispeaker system.

crossover spiral— *See* Leadover Groove.

cross talk—Interference caused by stray electromagnetic or electrostatic coupling of energy from one circuit to another.

crystal cutter—A disc cutter in which mechanical displacement of the recording stylus is obtained by applying the recording signal to a piezoelectric crystal.

crystal diode—A two-electrode semiconductor device that uses the rectifying properties of a pn junction. Also called crystal rectifier, diode, and semiconductor diode.

crystal headphones—Headphones whose active elements are Rochelle-salt crystals or similar crystals with piezoelectric properties.

crystal loudspeaker—A speaker in which piezoelectric action is used to produce mechanical dis-

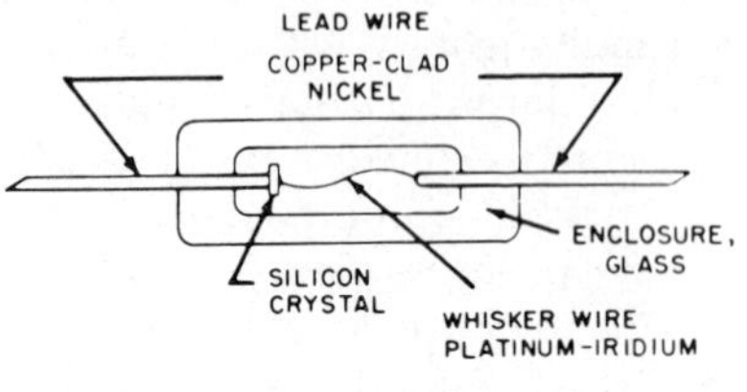

Crystal diode.

placement. Also called piezoelectric loudspeaker.

crystal microphone—Also called piezoelectric microphone. A microphone that uses a mechanical-to-electric transducer of the piezoelectric type.

crystal pickup—Also called piezoelectric pickup. A phonograph pickup that uses a transducer element having piezoelectric properties.

crystal speaker—A speaker in which piezoelectric action is used to produce mechanical displacement. Also called piezoelectric speaker.

crystal transducer— *See* Piezoelectric Transducer.

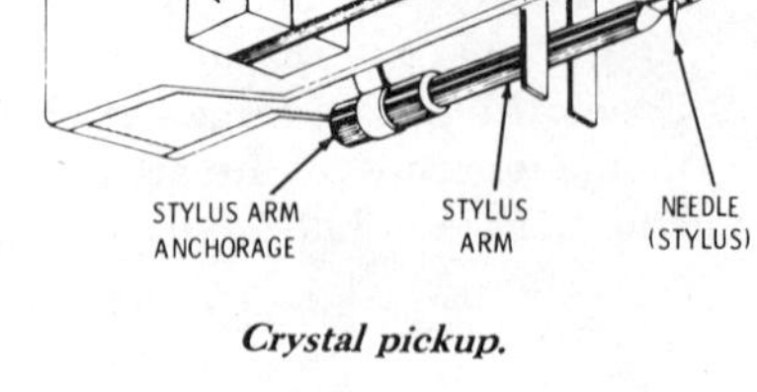

Crystal pickup.

cutoff—1. The frequency above or below which a selective circuit fails to respond. 2. The frequency of transmission at which the loss exceeds by 10 decibels the loss observed at 1000 hertz.

cutoff frequency—1. The frequency at which the gain of an amplifier falls below .707 times the maximum gain. 2. The frequency that marks the edge of the passband of a filter and the beginning of the transition to the stopband.

cutter—Also called mechanical recording head. An electromechanical transducer that transforms an electric input into a mechanical output (for example, the mechanical motions that a cutting stylus inscribes into a recording medium).

cutting rate—In disc recording, the number of lines per inch the lead screw moves the cutting-head carriage across the face of a recording blank. Standard rates are 96, 104, 112, 128, 136 and greater, in multiples of eight lines per inch. For microgroove recordings, 200 to 300 lines per inch are used.

cycle—The change of an alternating wave from zero to a negative peak, to zero, to a positive peak, and back to zero. The number of cycles per second (hertz) is called the frequency. *See also* Alternation.)

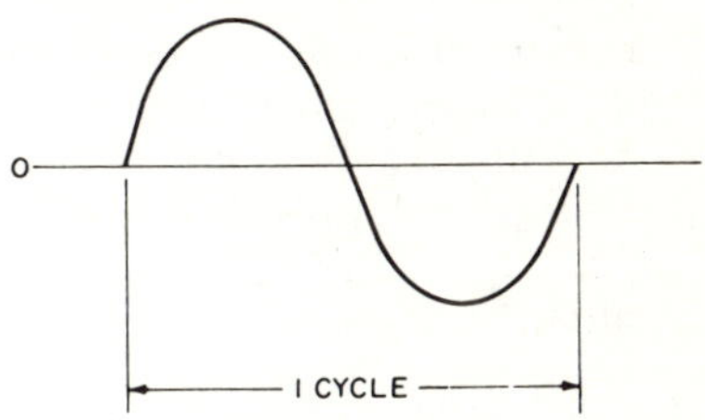

Cycle.

D

damped natural frequency—The frequency at which a system with a single degree of freedom will oscillate, in the presence of damping, after momentary displacement from the rest position by a transient force.

damped oscillation—The oscillation that occurs when the amplitude of the oscillating quantity decreases with time. If the rate of decrease can be expressed as a mathematical function, the name of the function describes the damping. Thus, if the rate of decrease is expressed as a negative exponential, the system is said to be exponentially damped.

damped waves—Waves in which successive cycles at the source progressively diminish in amplitude.

dampen—1. To diminish progressively in amplitude; usually said of waves or oscillations. 2. To deaden vibrations.

damping—1. Reduction of energy in a mechanical or electrical system by absorption or radiation. 2. Act of reducing the amplitude of oscillations of an oscillatory system, hindering or preventing oscillation or vibration, or diminishing the sharpness of resonance of the natural frequency of a system. 3. The dissipation of kinetic energy in a system by a controlled, energy-absorbing medium.

damping coefficient—The ratio of actual damping to critical damping.

damping factor—1. Numerical quantity indicating the ability of an amplifier to operate a speaker properly. Values over 4 are usually considered satisfactory. 2. The ratio of rated load impedance to the internal impedance of an amplifier.

damping ratio—The ratio of the degree of actual damping to the degree of damping required for critical damping. May be affected by changes in ambient temperature.

dB—Abbreviation for decibel.

dBa—Abbreviation for decibels adjusted. Used in conjunction with noise measurements. The reference level is –90 dBm, and the adjustment depends on the frequency-band weighting characteristics of the measuring device.

dBk—Decibels referred to 1 kilowatt.

dBm—Abbreviation for decibels above (or below) one milliwatt. A quantity of power expressed in terms of its ratio to 1 milliwatt.

dBRAP—Decibels above reference acoustical power, which is defined as 10^{-16} watt.

dBRN—Decibels above reference noise. Reference noise level is –90 dBm at 1000 hertz.

dBV—The increase or decrease in voltage, independent of impedance levels.

dBW—Decibels referred to 1 watt.

dc—Abbreviation for direct current.

dc amplifier— *See* Direct-Current Amplifier.

dc noise—Noise that arises when reproducing a tape that has been nonuniformly magnetized by energizing the record head with direct current, either in the presence or absence of bias. This noise has pronounced long-wavelength components that can be as much as 20 dB higher than those obtained from a bulk-erased tape.

dead end—In a sound studio, the end with the greater sound-absorbing characteristic.

dead room—A room for testing the acoustic efficiency or range of electroacoustic devices such as speakers and microphones. The room is designed with an absolute minimum of sound reflection, and no two dimensions of the room are the same.

decibel (dB)—The unit for comparing relative levels of sound waves or of voltage or power signals in amplifiers. The formula for voltage comparisons is:

$$dB = 20 \log_{10} \frac{E_1}{E_2}$$

The formula for power comparisons is:

$$dB = 10 \log_{10} \frac{P_1}{P_2}$$

See the accompanying table.

deck (tapedeck)— The tape-transporting mechanism, including the heads, drive, and tape-motion controls.

decoder—A device for recovering information that has been encoded. An example is the decoder circuit used in fm stereo receivers to obtain the second channel information.

de-emphasis—Also called post-emphasis or post-equalization. Introduction of a frequency-response characteristic that is complementary to that introduced in pre-emphasis.

de-emphasis network—A network inserted into a system to restore the pre-emphasized frequency spectrum to its original form.

delay—1. The time required for a signal to pass through a device or conductor. 2. The time interval

dB	Current or Voltage Ratio		Power Ratio	
	Gain	Loss	Gain	Loss
0	1.000	1.0000	1.000	1.0000
.1	1.012	.9886	1.023	.9772
.2	1.023	.9772	1.047	.9550
.3	1.035	.9661	1.072	.9333
.4	1.047	.9550	1.096	.9120
.5	1.059	.9441	1.122	.8913
.6	1.072	.9333	1.148	.8710
.7	1.084	.9226	1.175	.8511
.8	1.096	.9120	1.202	.8318
.9	1.109	.9016	1.230	.8128
1.0	1.122	.8913	1.259	.7943
1.1	1.135	.8810	1.288	.7762
1.2	1.148	.8710	1.318	.7586
1.3	1.161	.8610	1.349	.7413
1.4	1.175	.8511	1.380	.7244
1.5	1.189	.8414	1.413	.7079
1.6	1.202	.8318	1.445	.6918
1.7	1.216	.8222	1.479	.6761
1.8	1.230	.8128	1.514	.6607
1.9	1.245	.8035	1.549	.6457
2.0	1.259	.7943	1.585	.6310
2.1	1.274	.7852	1.622	.6166
2.2	1.288	.7762	1.660	.6026
2.3	1.303	.7674	1.698	.5888
2.4	1.318	.7586	1.738	.5754
2.5	1.334	.7499	1.778	.5623
2.6	1.349	.7413	1.820	.5495
2.7	1.365	.7328	1.862	.5370
2.8	1.380	.7244	1.905	.5248
2.9	1.396	.7161	1.950	.5129
3.0	1.413	.7079	1.995	.5012
3.1	1.429	.6998	2.042	.4898
3.2	1.445	.6918	2.089	.4786
3.3	1.462	.6839	2.138	.4677
3.4	1.479	.6761	2.188	.4571
3.5	1.496	.6683	2.239	.4467
3.6	1.514	.6607	2.291	.4365
3.7	1.531	.6531	2.344	.4266
3.8	1.549	.6457	2.399	.4169
3.9	1.567	.6383	2.455	.4074
4.0	1.585	.6310	2.512	.3981
4.1	1.603	.6237	2.570	.3890
4.2	1.622	.6166	2.630	.3802
4.3	1.641	.6095	2.692	.3715
4.4	1.660	.6026	2.754	.3631
4.5	1.679	.5957	2.818	.3548
4.6	1.698	.5888	2.884	.3467
4.7	1.718	.5821	2.951	.3388
4.8	1.738	.5754	3.020	.3311
4.9	1.758	.5689	3.090	.3236
5.0	1.778	.5623	3.162	.3162
5.1	1.799	.5559	3.236	.3090
5.2	1.820	.5495	3.311	.3020
5.3	1.841	.5433	3.388	.2951
5.4	1.862	.5370	3.467	.2884
5.5	1.884	.5309	3.548	.2818
5.6	1.905	.5248	3.631	.2754
5.7	1.928	.5188	3.715	.2692
5.8	1.950	.5129	3.802	.2630
5.9	1.972	.5070	3.890	.2570
6.0	1.995	.5012	3.981	.2512
6.1	2.018	.4955	4.074	.2455
6.2	2.042	.4898	4.169	.2399
6.3	2.065	.4842	4.266	.2344
6.4	2.089	.4786	4.365	.2291
6.5	2.113	.4732	4.467	.2239
6.6	2.138	.4677	4.571	.2188
6.7	2.163	.4624	4.677	.2138
6.8	2.188	.4571	4.786	.2089
6.9	2.213	.4519	4.898	.2042
7.0	2.239	.4467	5.012	.1995
7.1	2.265	.4416	5.129	.1950
7.2	2.291	.4365	5.248	.1905
7.3	2.317	.4315	5.370	.1862
7.4	2.344	.4266	5.495	.1820
7.5	2.371	.4217	5.623	.1778
7.6	2.399	.4169	5.754	.1738
7.7	2.427	.4121	5.888	.1698
7.8	2.455	.4074	6.026	.1660
7.9	2.483	.4027	6.166	.1622
8.0	2.512	.3981	6.310	.1585
8.1	2.541	.3936	6.457	.1549
8.2	2.570	.3890	6.607	.1514
8.3	2.600	.3846	6.761	.1479
8.4	2.630	.3802	6.918	.1445
8.5	2.661	.3758	7.079	.1413
8.6	2.692	.3715	7.244	.1380
8.7	2.723	.3673	7.413	.1349
8.8	2.754	.3631	7.586	.1318
8.9	2.786	.3589	7.762	.1288
9.0	2.818	.3548	7.943	.1259
9.1	2.851	.3508	8.128	.1230
9.2	2.884	.3467	8.318	.1202
9.3	2.917	.3428	8.511	.1175
9.4	2.951	.3388	8.710	.1148
9.5	2.985	.3350	8.913	.1122
9.6	3.020	.3311	9.120	.1096
9.7	3.055	.3273	9.333	.1072
9.8	3.090	.3236	9.550	.1047
9.9	3.126	.3199	9.772	.1023
10.0	3.162	.3162	10.000	.1000
10.1	3.199	.3126	10.23	.09772
10.2	3.236	.3090	10.47	.09550
10.3	3.273	.3055	10.72	.09333
10.4	3.311	.3020	10.96	.09120
10.5	3.350	.2985	11.22	.08913
10.6	3.388	.2951	11.48	.08710
10.7	3.428	.2917	11.75	.08511
10.8	3.467	.2884	12.02	.08318
10.9	3.508	.2851	12.30	.08128

Decibel table (0 to 10.9 dB).

dB	Current or Voltage Ratio		Power Ratio	
	Gain	Loss	Gain	Loss
11.0	3.548	.2818	12.59	.07943
11.1	3.589	.2786	12.88	.07762
11.2	3.631	.2754	13.18	.07586
11.3	3.673	.2723	13.49	.07413
11.4	3.715	.2692	13.80	.07244
11.5	3.758	.2661	14.13	.07079
11.6	3.802	.2630	14.45	.06918
11.7	3.846	.2600	14.79	.06761
11.8	3.890	.2570	15.14	.06607
11.9	3.936	.2541	15.49	.06457
12.0	3.981	.2512	15.85	.06310
12.1	4.027	.2483	16.22	.06166
12.2	4.074	.2455	16.60	.06026
12.3	4.121	.2427	16.98	.05888
12.4	4.169	.2399	17.38	.05754
12.5	4.217	.2371	17.78	.05623
12.6	4.266	.2344	18.20	.05495
12.7	4.315	.2317	18.62	.05370
12.8	4.365	.2291	19.05	.05248
12.9	4.416	.2265	19.50	.05129
13.0	4.467	.2239	19.95	.05012
13.1	4.519	.2213	20.42	.04898
13.2	4.571	.2188	20.89	.04786
13.3	4.624	.2163	21.38	.04677
13.4	4.677	.2138	21.88	.04571
13.5	4.732	.2113	22.39	.04467
13.6	4.786	.2089	22.91	.04365
13.7	4.842	.2065	23.44	.04266
13.8	4.898	.2042	23.99	.04169
13.9	4.955	.2018	24.55	.04074
14.0	5.012	.1995	25.12	.03981
14.1	5.070	.1972	25.70	.03890
14.2	5.129	.1950	26.30	.03802
14.3	5.188	.1928	26.92	.03715
14.4	5.248	.1905	27.54	.03631
14.5	5.309	.1884	28.18	.03548
14.6	5.370	.1862	28.84	.03467
14.7	5.433	.1841	29.51	.03388
14.8	5.495	.1820	30.20	.03311
14.9	5.559	.1799	30.90	.03236
15.0	5.623	.1778	31.62	.03162
15.1	5.689	.1758	32.36	.03090
15.2	5.754	.1738	33.11	.03020
15.3	5.821	.1718	33.88	.02951
15.4	5.888	.1698	34.67	.02884
15.5	5.957	.1679	35.48	.02818
15.6	6.026	.1660	36.31	.02754
15.7	6.095	.1641	37.15	.02692
15.8	6.166	.1622	38.02	.02630
15.9	6.237	.1603	38.90	.02570
16.0	6.310	.1585	39.81	.02512
16.1	6.383	.1567	40.74	.02455
16.2	6.457	.1549	41.69	.02399
16.3	6.531	.1531	42.66	.02344
16.4	6.607	.1514	43.65	.02291
16.5	6.683	.1496	44.67	.02239
16.6	6.761	.1479	45.71	.02188
16.7	6.839	.1462	46.77	.02138
16.8	6.918	.1445	47.86	.02089
16.9	6.998	.1429	48.98	.02042
17.0	7.079	.1413	50.12	.01995
17.1	7.161	.1396	51.29	.01950
17.2	7.244	.1380	52.48	.01905
17.3	7.328	.1365	53.70	.01862
17.4	7.413	.1349	54.95	.01820
17.5	7.499	.1334	56.23	.01778
17.6	7.586	.1318	57.54	.01738
17.7	7.674	.1303	58.88	.01698
17.8	7.762	.1288	60.26	.01660
17.9	7.852	.1274	61.66	.01622
18.0	7.943	.1259	63.10	.01585
18.1	8.035	.1245	64.57	.01549
18.2	8.128	.1230	66.07	.01514
18.3	8.222	.1216	67.61	.01479
18.4	8.318	.1202	69.18	.01445
18.5	8.414	.1189	70.79	.01413
18.6	8.511	.1175	72.44	.01380
18.7	8.610	.1161	74.13	.01349
18.8	8.710	.1148	75.86	.01318
18.9	8.810	.1135	77.62	.01288
19.0	8.913	.1122	79.43	.01259
19.1	9.016	.1109	81.28	.01230
19.2	9.120	.1096	83.18	.01202
19.3	9.226	.1084	85.11	.01175
19.4	9.333	.1072	87.10	.01148
19.5	9.441	.1059	89.13	.01122
19.6	9.550	.1047	91.20	.01096
19.7	9.661	.1035	93.33	.01072
19.8	9.772	.1023	95.50	.01047
19.9	9.886	.1012	97.72	.01023

Note: For values from 20 to 100 dB, see next page.

Decibel table (11 to 19.9 dB).

dB	Current or Voltage Ratio		Power Ratio
	Gain	Loss	
20.0	10.00 Use the same numbers as 0-20 dB, but shift point one step to the right. Thus since 10 dB = 3.162, 30 dB = 31.62.	.1000 Use the same numbers as 0-20 dB, but shift point one step to the left. Thus since 10 dB = .3162, 30 dB = .03162.	10.00 This column repeats every 10 dB instead of every 20 dB.
40.0	100 Use the same numbers of 0-20 dB, but shift point two steps to the right. Thus since 10 dB = 3.162, 50 dB = 316.2.	.01 Use the same numbers as 0-20 dB, but shift point two steps to the left. Thus since 10 dB = .3162, 50 dB = .003162.	20.00 This column repeats every 10 dB instead of every 20 dB.
60.0	1000 Use the same numbers as 0-20 dB, but shift point three steps to the right. Thus since 10 dB = 3.162, 70 dB = 3162.	.001 Use the same numbers as 0-20 dB, but shift point three to the left. Thus since 10 dB = .3162, 70 dB = .0003162.	30.00 This column repeats every 10 dB instead of every 20 dB.
80.0	10,000 Use the same numbers as 0-20 dB, but shift point four steps to the right. Thus since 10 dB = 3.162, 90 dB = 31620.	.0001 Use the same numbers as 0-20 dB, but shift point four steps to the left. Thus since 10 dB = .3162, 90 dB = .00003162.	40.00 This column repeats every 10 dB instead of every 20 dB.
100.0	100,000	.00001	50.00

Decibel table (20 to 100 dB).

between the instants at which any designated point in a wave passes any two designated points of a transmission circuit.

delay circuit—A circuit that delays the passage of a pulse or signal from one part of a circuit to another.

delay equalizer—1. A device that adds delay at certain frequencies to a circuit in a way to reduce the delay distortion. 2. A corrective network that is designed to make the phase delay or envelope delay of a circuit or system substantially constant over a desired frequency range. 3. A network that introduces an amount of phase shift in the circuit at all frequencies within the desired band.

delay line—A real or artificial transmission line or equivalent device designed to delay a signal or wave for a predetermined length of time.

delay time—The amount of time one signal is behind or lags another.

demagnetization—Partial or complete reduction of residual magnetism.

demagnetization curve—In the second quadrant of a hysteresis loop, the portion that lies between the residual induction point, B_r, and the coercive point, H_c.

demagnetization effect—A decrease in internal magnetic field caused by uncompensated magnetic poles at the surface of a sample.

demagnetizer—A device for removing the magnetism that may build up in a recording or playback head.

demagnetizing force—A magnetizing force applied in such a direction that it reduces the residual induction in a magnetized body.

demodulation—Also called detection. The operation on a previously modulated wave in such a way that it will have substantially the same characteristics as the original modulating wave.

demodulator—A device that operates on a carrier wave to recover the wave with which the carrier was originally modulated.

demultiplexer—A device used to separate two or more signals that were previously combined by a compatible multiplexer and are transmitted over a single channel.

demultiplexing circuit—A circuit that is used to separate signals that have been combined for transmission by multiplex.

destaticization—Treatment of a material to minimize the accumulation of static electricity and, as a result, the amount of dust that adheres to the material because of such static charges.

detection— *See* Demodulation.

detector—A device for effecting the process of detection or demodulation.

deviation—In frequency modulation, the difference between the instantaneous frequency of the modulating signal and the carrier frequency.

diamagnetic—Term applied to a substance with a negative magnetic susceptibility.

diamagnetic material—A material in which the intensity of magnetism is negative. There is no known material in which this effect has more than a very feeble intensity. Bismuth is the leading example of materials of this class.

diameter equalization—Variable equalization applied to a disc-recording circuit to compensate for high-frequency losses that would otherwise occur as the cutting head moves near the center of the record.

diamond stylus—A phonograph stylus with a shaped and polished diamond as its point.

diapason—The unique fundamental tone color of organ music.

diaphragm—A flexible membrane used in various electroacoustic transducers for producing audio-frequency vibrations when actuated by electric impulses, or, conversely, for producing electric impulses when actuated by audio-frequency vibrations.

difference channel—In a stereophonic sound system, an audio channel that handles the difference between the signals in the right and left channels.

differential microphone— *See* Double-Button Carbon Microphone.

diffracted wave—A radio, sound, or light wave that has struck an object and been bent or deflected, other than by reflection or refraction.

diffraction—The phenomenon whereby waves traveling in straight paths bend around an obstacle.

diffused sound—Sound that has uniform energy density, meaning that the energy flux is equal in all parts of a given region.

DIN— *Deutsche Industrie Normen* (German Industrial Standards). In audio components, the German standard of plugs, sockets, connectors, etc.

diode—1. An electron tube having two electrodes: a cathode and an anode. 2. *See* Crystal Diode.

direct current—Abbreviated dc. Current that is essentially constant in amplitude and flowing in one direction only. (Small, regularly occurring variations in the amplitude of a direct current are called an ac component of the dc.)

direct-current amplifier— Also called dc amplifier. An amplifier capable of boosting dc voltages. Resistive coupling only is generally employed between stages, but sometimes will be combined with other forms.

direct-current erasing head—In magnetic recording, a head that uses direct current to produce the magnetic field required for erasure. Direct current erasing subjects the medium to a unidirectional field. Such a medium is therefore in a different magnetic state than one erased by alternating current.

directional characteristic— The output of a sound source shown as a function of the angle in any given direction as compared with the zero axis of the system. The characteristics are usually plotted on a polar graph.

directional microphone—A microphone the response of which varies significantly with the direction of sound. (*See also* Unidirectional, Bidirectional, and Semidirectional Microphone.)

direct piezoelectricity—A name sometimes given to the piezoelectric effect in which an electric charge develops on a crystal when it is subjected to mechanical stress.

direct-radiator speaker—A speaker in which the radiating element acts directly on the air instead of delivering its output to some intervening element such as a horn.

direct-recording magnetic tape—A method of recording using a high-frequency bias in which the electrical input signal is applied to the recording head without alteration.

direct sound wave—A wave emitted in an enclosure by a source prior to the time it has undergone its first reflection from a boundary of the enclosure. Frequently a sound wave is said to be direct if it contains reflections that have occurred from surfaces within about 0.05 second after the sound was first emitted.

disc—1. A phonograph record. 2. The blank used in a disc recorder.

disc recorder—A recording device in which the sound information is mechanically impressed onto a disc—as opposed to a tape recorder, which impresses the sound magnetically on a tape.

discrete—A term used in the nomenclature of quadraphonic sound. In its strictest sense, it means that the four program channels remain independent of each other throughout the recording and playback process. A more liberal interpretation of the term is that the entire process finally leads to the recovery of fully separated L_f, R_f, and L_b, and R_b signals.

disc storage—1. A storage device that uses magnetic recording on flat rotating discs. 2. The storage of data on the surface of magnetic discs.

displacement transducer—A device that converts the energy involved in mechanical displacement into electrical energy. The amount of output voltage is determined by the amount of displacement.

dissymmetrical transducer—Also called dissymmetrical network. A transducer with unequal input and output image impedances.

distortion—Any element that alters the faithful reproduction of the original sound. Distortion is further classified as nonlinear, frequency, or phase, depending upon how the waveform is affected.

dividing network—Also called speaker dividing network and crossover network. A frequency-selective network that divides the audio-frequency spectrum into two or more parts to be fed to separate devices such as amplifiers or speakers.

Dolby A-type noise-reduction system—A system for reducing noise in commercial recording and reproduction. Based on compression and expansion, with program material divided into four frequency bands, each treated individually.

Dolby B-type noise-reduction system—A system for noise reduction in recording and playback, with wide application to cassette units. The system uses successive compression and expansion, as with the A-system, but deals with frequencies above 500 Hz, and automatically reduces the bandwidth of its operation as signal level increases.

"Dolbyized" tape—Prerecorded tape that has been recorded using the Dolby noise-reduction system. Such tape is properly reproduced on machines equipped with Dolby playback circuitry. If played without a Dolby circuit, it may sound excessively brilliant.

Dolby system—A noise reduction system that supplies signal boost on low-level signals during recording and a corresponding cut during playback.

Doppler cabinet—A speaker cabinet for producing a vibrato effect. Either the speaker or a baffle board is rotated or moved to produce continuous change in the length of the sound path.

Doppler effect—The change in the frequency of a sound caused by movement of the source, movement of the observer, or both. The same name is also applied to a similar phenomenon with light waves.

double-button carbon microphone—Also called differential microphone. A microphone with two carbon-resistance elements or buttons, one on each side of a central diaphragm. They are connected in parallel to the current source to give twice the resistance change obtainable with a single button.

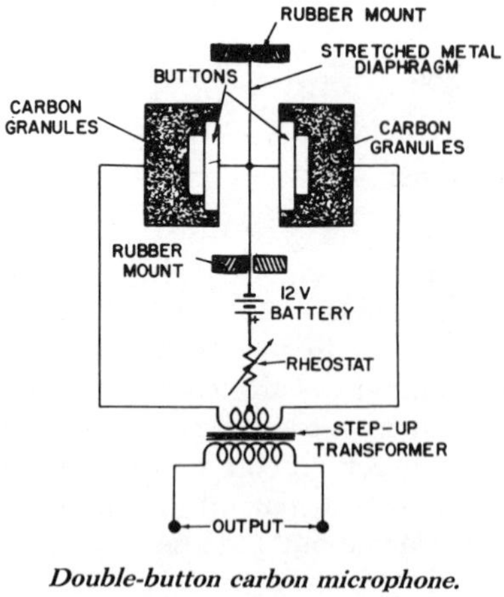

Double-button carbon microphone.

doubling—The generation of large amounts of second-harmonic distortion by nonlinear motion of a speaker cone.

drag angle—A stylus cutting angle of less than 90° to the surface of the record. So called because the stylus drags over the surface instead of digging in.

drive pin—In disc recording, a pin mounted a short distance from the center pin or turntable shaft and used to prevent the disc from slipping on the turntable.

drone cone—An undriven speaker cone mounted in a bass-reflex enclosure.

dual-groove record— *See* Cook System.

dual pickup— *See* Turnover Pickup.

dual-track recorder—Sometimes called a half-track recorder. The recording head covers only half the width of the tape, making it possible to record on one half of the tape and then, by turning the reels over, to record on the unused half.

dubbing—The addition of sound to a prerecorded tape or disc.

dynamic microphone— *See* Moving-Coil Microphone.

dynamic noise suppressor—An audio filter designed to automatically adjust the bandpass with changes in the signal level. At low signal levels filtering is highest, and at high signal levels all filtering action is removed.

dynamic pickup—A phonograph pickup that produces electrical output as the result of the motion of a conductor in a magnetic field.

dynamic range—The ratio between the softest and the loudest sounds that an amplifier or recorder can reproduce within acceptable distortion limits.

dynamic reproducer— *See* Moving-Coil Pickup.

dynamic speaker— *See* Moving-Coil Speaker.

dyne—The force that produces an acceleration of one centimeter per second per second on one gram of mass.

dyne per square centimeter—The unit of sound pressure. One dyne per square centimeter was originally called a bar in acoustics, but the full expression is used now because the bar is defined differently in other applications. Also called microbar.

E

earphone—Also called receiver. An electroacoustic transducer intended to be placed in or over the ear.

eccentric groove—Also called eccentric circle. An off-center locked groove for actuating the trip mechanism of an automatic record changer at the end of a record.

echo chamber—A reverberant room or enclosure used for adding hollow effects or actual echoes to radio or television programs.

editing—Changing the sequence or content of a tape recording by deleting, rearranging, or adding material.

effective acoustic center—The point from which the spherically divergent sound waves from an acoustic generator appear to diverge. Also called apparent source.

effective sound pressure—The root-mean-square value of the instantaneous sound pressure at one point over a complete cycle. The unit is the dyne per square centimeter.

effective value—Also called the rms (root-mean-square) value. The value of alternating current that will produce the same amount of heat in a resistance as the corresponding value of direct current. For a sine wave, the effective value is 0.707 times the peak value.

efficiency—Ratio of the useful output of a device to the input to the device. For practical devices, this is always a number less than 1. Usually expressed in percent, found by multiplying the ratio just described by 100.

elastic limit—The maximum stress a solid can endure and still return to its unstrained dimensions when the stress is removed.

elastic wave—A pure acoustic wave.

electret microphone—A microphone based on the principles of a condenser microphone. A thin metallized plastic film permanently polarized in the manner of an electret is used as one plate of the condenser. This eliminates the need for a high-voltage supply as used to polarize the diaphragm of a standard condenser microphone.

electrical angle—A quantity that specifies a particular instant in a cycle of alternating current. One cycle is considered to be 360°, a half cycle is 180°, etc.

electrical inertia—Inductance that opposes any changes in current through an inductor.

electrical load—A device (e.g., a speaker) comprising resistive or reactive components into which an amplifier, generator, etc., delivers power.

electric strain gauge—A device for measuring strain in structural members. The principle has been applied in one type of phonograph pickup. The stress applied to the stylus of the pickup by the record groove creates strain in the sensitive element in the pickup, resulting in a corresponding change of current through the pickup terminals.

electric transcription—In broadcasting, a disc recording of a message or a complete program.

electroacoustic—Pertaining to a device (e.g., a speaker or a microphone) that involves both electric current and sound-frequency pressures.

electroacoustic transducer—A device that receives signals from an electric system and delivers an output to an acoustic system, or vice versa. A speaker is an example of the first case and a microphone is an example of the second.

electrodynamic—Pertaining to electric current, electricity in motion, and the actions and effects of magnetism and induction.

electrodynamic braking— A method of stopping a tape-deck motor gently by the application of a predetermined dc voltage to the motor.

electrodynamic speaker—A speaker containing an electromagnet called the field coil, through which a direct current flows.

electrolytic capacitor—A capacitor consisting of two electrodes separated by a dielectric: one electrode, the anode, carries the dielectric as a very thin oxide film; the other electrode consists of an electrolyte. This type of capacitor permits a relatively high capacitance in a small physical size.

electromagnet—A solenoid with current passing through it. An iron core is usually included to increase the field strength. When the current is interrupted, the magnetism disappears.

electromagnetic—Dependent on electricity for its magnetic properties.

electromagnetic field—1. The magnetic field of influence produced around a conductor by the

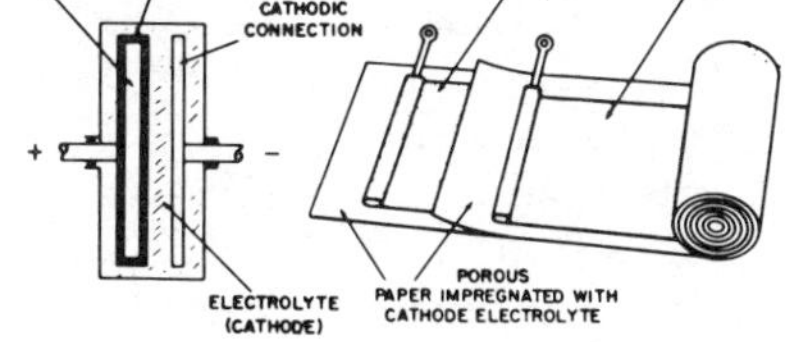

Electrolytic capacitor.

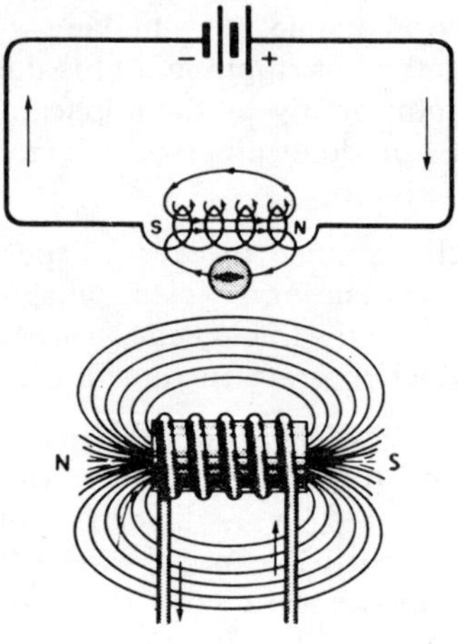

Electromagnet.

current flowing through it. 2. A rapidly moving electric field and its associated magnetic field.

electromagnetic induction—The phenomenon whereby voltage is produced in a coil as the number of magnetic lines of force (flux linkages) passing through the coil changes.

electromagnetic inertia—1. The opposition to change of current in an electric circuit on application or removal of the source voltage. 2. The property of self-induction.

electromagnetic radiation—A form of radiated energy characterized by transversely oscillating electric and magnetic fields and propagated at velocity "c" in free space. At a sufficient distance from the radiating source, the electric-field vector and magnetic-field vector are at right angles to each other, forming a right-handed (coordinate) system. In an ionized medium, a longitudinal component may be present.

electromagnetic-type microphones—Microphones in which the voltages are varied by an electromagnet (namely, ribbon or velocity, dynamic or moving-coil, and reluctance or moving-vane microphones).

electromechanical—Any device using electrical energy to produce mechanical movement.

electromechanical transducer—A device that transforms electrical energy into mechanical energy or vice versa. A speaker is an example of the first type and a microphone is an example of the second type.

electromotive force—Abbreviated emf. The work done on moving electric charges to maintain a continuous difference of potential. The unit of measurement is the volt.

electronic microphone—A device that depends for its operation on the generation of a voltage by the motion of one of the electrodes in a special electron tube.

electronic music—Music made by electronic synthesis and processing of musical tones or by electronic alteration of natural sounds.

electronic organ—An electronic keyboard instrument patterned after the pipe organ but having all tones and tone variations such as vibrato, tremolo, etc., produced by electronic circuits instead of by pipes.

electrostatic microphone—Also called capacitor or condenser microphone. A microphone that depends for its operation on variations in its electrostatic capacitance.

electrostatic speaker—Also called capacitor or condenser speaker. A speaker in which the mechanical forces are produced by the action of electrostatic fields on movable elements.

electrostatic transducer—A transducer that consists of a capacitor, at least one plate of which can be set into vibration. Its operation depends on the attractive and repulsive forces between electrically charged bodies.

electrostatic tweeter—A speaker that is essentially a capacitor with one of the electrodes being movable. A polarizing voltage is applied between electrodes and the audio signal is superimposed on this voltage, alternately opposing and reinforcing it. Electrode movement occurs because of electrostatic forces.

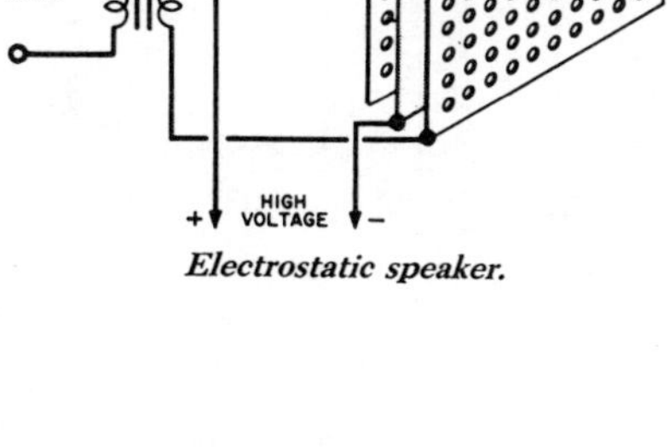

Electrostatic speaker.

electrostriction — Dimensional change caused by the application of an electric field to any dielectric material.

embossed-groove recording—A method of recording sounds on discs or film strips by embossing sound grooves with a blunt stylus rather than cutting into them with a sharp stylus.

embossing stylus—A recording stylus with a rounded tip that forms a groove in the recording medium by merely displacing the material instead of cutting it out completely.

emf—Abbreviation for electromotive force.

emitter—1. In a junction transistor, the junction that is usually forward biased. 2. That element of a transistor analogous to the cathode of a vacuum tube.

emitter follower—A common-collector transistor amplifier similar in operation to a vacuum-tube cathode follower.

Embossed-groove recording.

enclosure—A housing for any electrical or electronic device, but normally for a speaker.

equalization—The process of changing the frequency response of a circuit by the use of appropriate networks that compensate for some loss or distortion incurred at some stage in the recording or reproducing system.

equalizer—A passive device designed to compensate for an undesired amplitude-frequency or phase-frequency characteristic of a system or component.

equivalent absorption—The rate at which a surface will absorb sound energy, expressed in sabins. Defined as the area of a perfect absorption surface that will absorb the same sound energy as the given object under the same conditions.

equivalent loudness—The intensity level of a sound relative to some arbitrary reference intensity, such as a 1000-hertz pure tone that is judged by the listeners to be equivalent in loudness.

equivalent noise pressure— *See* Transducer Equivalent Noise Pressure.

erase head—That part of a tape recorder that is used to erase unwanted program material on the tape. Usually it is an electromagnet operated with a supersonic electronic signal, but sometimes can be a permanent magnet.

erase oscillator—In a magnetic recorder, an oscillator that provides the high-frequency signal used for erasing a magnetic tape.

erasing head—A head used to remove previously recorded material. It may also be used to precondition tapes for recording purposes.

erasure—The process by which a signal recorded on a tape is removed and the tape made ready for rerecording. This may be accomplished by ac erasure, where the tape is demagnetized by an alternating field that is reduced in amplitude from an ini-

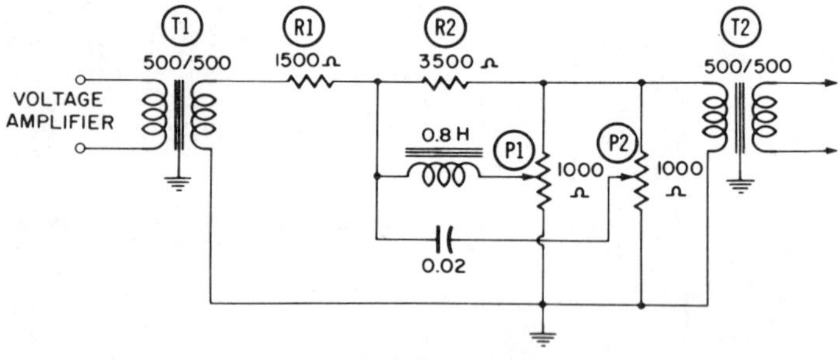

Equalizer circuit.

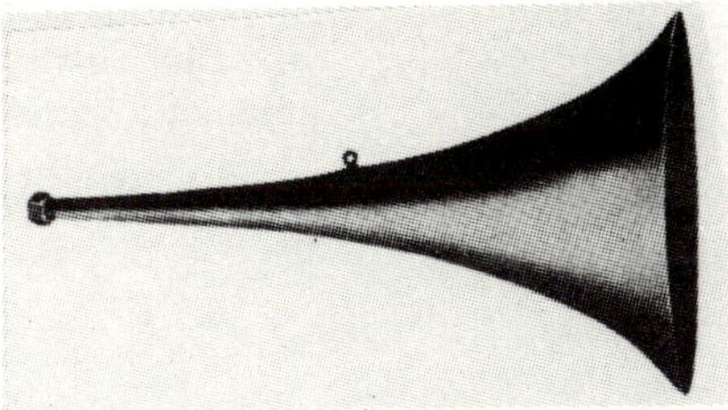

Exponential horn. (Courtesy Bell Telephone Labs)

tially high value, or by dc erasure, where the tape is saturated by applying a primarily unidirectional field.

erg—The absolute centimeter-gram-second unit of energy and work. The work done when a force of one dyne is applied through a distance of one centimeter.

E-vector—A vector representing the electric field of an electromagnetic wave. In free space, it is perpendicular to the direction of propagation.

even harmonic—Any harmonic that is an even multiple (2, 4, 6, etc.) of the fundamental frequency.

excess sound pressure—The total instantaneous pressure at a point in a medium containing sound waves, minus the static pressure when no sound waves are present. The unit is the dyne per square centimeter.

excited-field speaker—A speaker in which the steady magnetic field is produced by an electromagnet.

exponential damping—Damping that follows an exponential law.

exponential decay—The decay of some measurable quantity at an exponential rate.

exponential horn—A horn with a cross-sectional area that changes exponentially with axial distance.

expression control—In an organ, the foot control that regulates the overall volume.

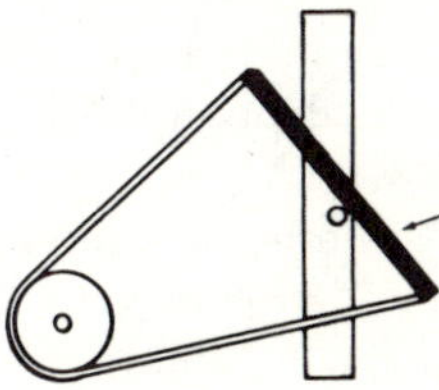

Expression control.

extended octaves—In an organ, tones above or below the notes on the regular keyboard that can be sounded only when certain couplers are on.

extended play—Abbreviated ep. A 45-rpm record on a 7-inch disc. It provides eight minutes of playing time, instead of the five minutes of a standard 45-rpm disc.

F

fade in—A gradual increase in signal strength in a sound or television channel.

fade out—A gradual decrease in signal strength in a sound or television channel.

fader—A multiple-unit control used in radio or sound systems for gradual changeover from one microphone or audio channel to another; in television, from one camera to another; and in motion-picture projection, from one projector to another.

fall-in—In a synchronous motor, the point at which synchronous speed is reached.

fall time—The time required for a waveform to decrease from 90% to 10% of its maximum amplitude.

farad—A unit of capacitance. The capacitance of a capacitor in which a charge of 1 coulomb produces a change of 1 volt in the potential difference between its terminals.

Faraday's law—Also called the law of electromagnetic induction. When a magnetic field cuts a conductor, or when a conductor cuts a magnetic field, an electric current will flow through the conductor if a closed path is provided.

fast forward—The provision on a tape recorder permitting tape to be run rapidly through the recorder in the play direction, usually for search or selection purposes.

fast groove—Also called fast spiral. In disc recording, an unmodulated spiral groove having a much greater pitch than the recorded grooves.

fast spiral— *See* Fast Groove.

feedback—In an amplifier system, the returning of a portion of the output signal to some preceding point in the system. If the phase of the signal fed back agrees with the phase of the signal at the point of application, it is called positive feedback; if it opposes the phase of the signal at that point, it is called negative feedback.

feedback amplifier—An amplifier that returns a portion of the output signal to an earlier point in the amplifier to modify its performance.

feedback cutter — An electromechanical transducer that is essentially a disc cutter, with the addition that it is equipped with an auxiliary feedback coil in the magnetic field. Signals exciting the cutter are induced into the feedback coil, the output of which is fed back in turn to the input of the cutter amplifier. The result is a substantially uniform frequency response.

feed reel—On a tape recorder, the reel that supplies the magnetic tape to the recording or playback head; hence, it is sometimes called the supply reel.

feedthrough—Accidental or unintentional transfer of signal from one track to another on a multitrack tape.

ferric oxide (Fe_2O_3)—A red, iron oxide coating used as the magnetic element of recording tapes.

fidelity—1. The accuracy with which the output of a system reproduces the characteristics of the input signal. 2. A measure of the exactness with which sound is duplicated or reproduced.

field magnet—An electromagnet or permanent magnet that produces a strong magnetic field in a magnetic transducer, generator, motor, or other electrical device.

field strength—Also called field intensity. The amount of magnetic flux produced at a particular point by an electromagnet or permanent magnet.

filler—In mechanical recording, the inert material of a record compound (as distinguished from the binder).

filter—A network of resistors, inductors, or capacitors that offers comparatively little opposition to certain frequencies or to direct current, while blocking or attenuating other frequencies.

filter attenuation—A loss of power through a filter as a result of reflection, radiation, or absorption in resistive materials. Usually expressed in decibels.

filter attenuation band—Also called filter stop band. A fre-

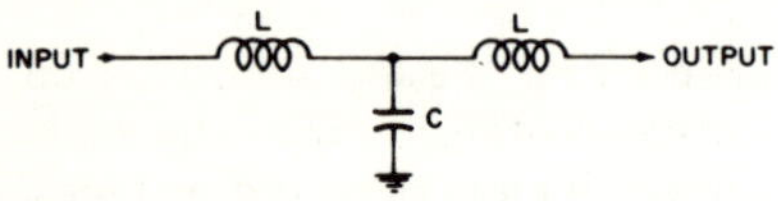

T-section.

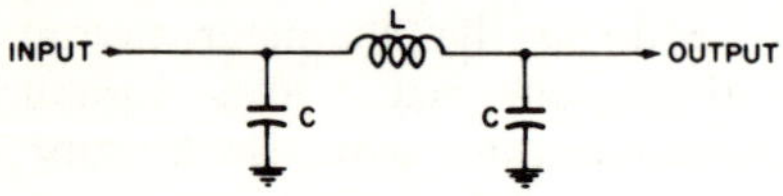

Pi-section.

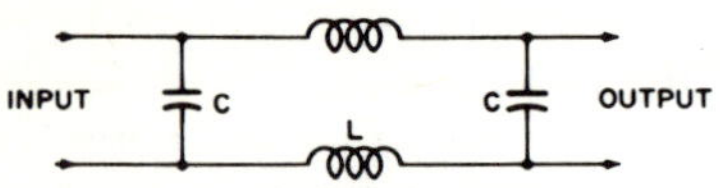

Ladder section.

Filters.

quency band in which there is attenuation besides any that might be due to dissipation.

filter stop band— *See* Filter Attenuation Band.

filter transmission band—Also called filter passband. A frequency band in which the attentuation constant is zero if dissipation is neglected. In other words, a frequency band of free transmission.

flame microphone—A microphone in which the action of sound waves on a flame changes the resistance between two electrodes in the flame.

flat frequency response—A system response that shows an essentially constant ratio of output to input for all frequencies within its accepted range. The maximum deviation that is permitted is usually stated as plus or minus a certain number of decibels.

flat response—A system is said to have a flat response when the amplitudes of input and output signals have a constant ratio to each other at all frequencies in question.

Fletcher-Munson curves—Also called equal-loudness contours. A group of curves showing intensity levels needed to produce sensations of equal loudness to the human ear at all audio frequencies for levels ranging from the threshold of hearing to the threshold of feeling.

flipover cartridge—A phonograph cartridge having separate needles for playing microgroove and standard records. It may be turned to bring the proper needle into playing position.

flock—Finely divided felt cemented to phonograph turntables, underneath microphone stands, or wherever a nonscratching surface is desired.

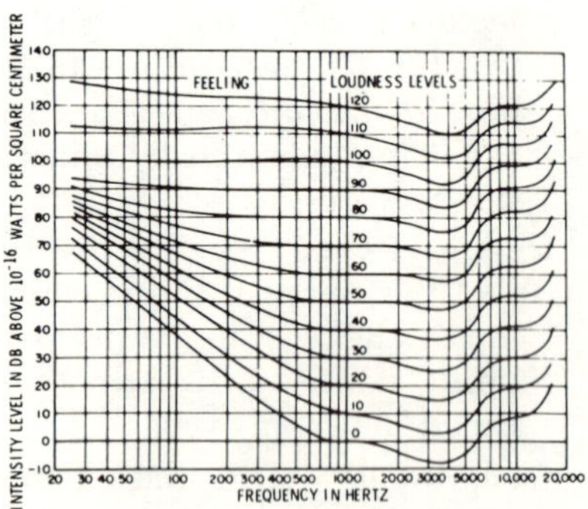

Fletcher-Munson curves.

flowed wax—A mechanical recording disc prepared by melting and flowing wax onto a metal disc.

flutter—Cyclic deviations from the normal steady motion of a turntable or tape transport. Also called "wow" and "drift," depending on cyclic rate. "Flutter" usually refers to relatively high rates (about 10 hertz). "Wow" refers to lower rates around one variation per turntable revolution. "Drift" refers to very slow rates around zero hertz and showing very little pattern of repetition.

flutter echo—1. A rapid succession of reflected pulses resulting from a single pulse. 2. A multiple echo in which the reflections occur in rapid succession. If periodic and audible, it is referred to as a musical echo.

flutter rate—The number of times per second the flutter repeats its cycle.

fm—Abbreviation for frequency modulation.

fm broadcast band—The band of radio frequencies extending from 88 to 108 MHz. These frequencies include those assigned to noncommercial educational broadcasting.

fm broadcast channel—A band of frequencies 200 kHz wide and designated by its center frequency. Channels for fm broadcast in the U.S. occupy the band from 88.1 MHz to 107.9 MHz in steps of 200 kHz.

fm multiplex— *See* FM Stereo.

fm stereo—Also called fm multiplex. A system by which fm radio

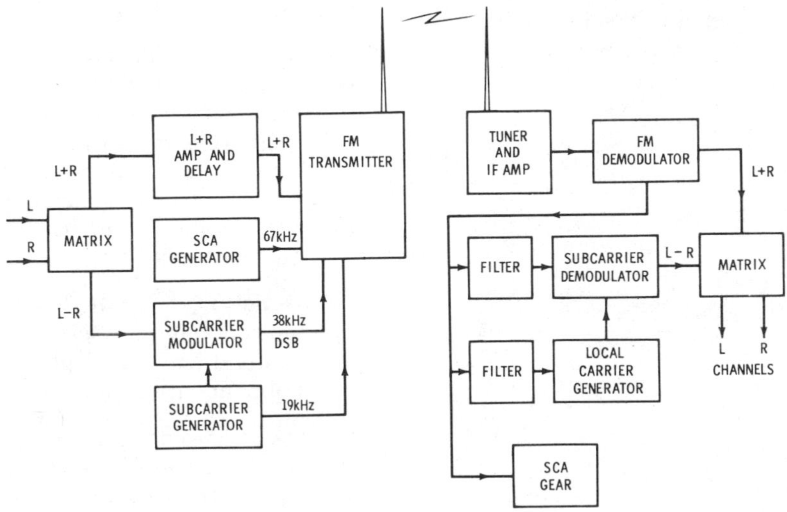

Fm stereophonic broadcast system.

stations are able to transmit stereophonic program material to specially designed receivers and which is at the same time compatible with monophonic equipment.

force—An agency or influence that, if applied to a free body, tends to produce acceleration or deformation of the body.

folded horn—An acoustic horn that is curled or folded to compress the horn length in as small a space as possible.

fork oscillator—An oscillator in which a tuning fork is the frequency-determining element.

fork tines—The projecting ends of a tuning fork. When vibrated, they produce a sound wave of constant frequency and low harmonic content.

formant—1. The particular frequency region in which the energy of a vowel sound is concentrated. 2. In an organ, a circuit for altering the tone quality of sound amplified by it. A formant filter is applied to the entire output of a manual, rather than to individual tones.

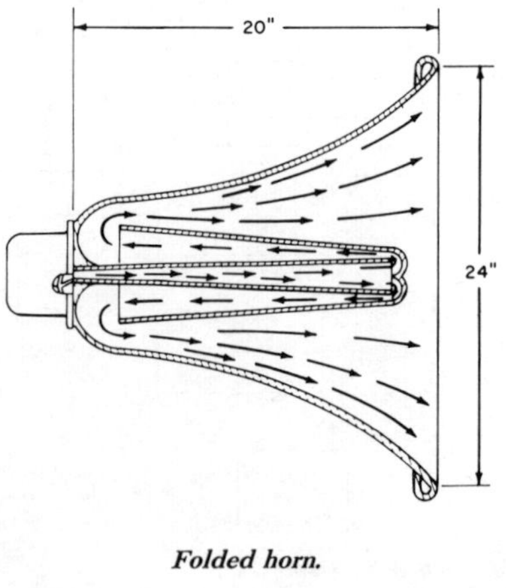

Folded horn.

formant filter—A waveshaping network used in an organ to modify the signal from the tone generator so it will assume the waveshape of the desired tone.

forty-five/forty-five—Also called the Westrex system. A system of disc recording in which signals originating from two microphones are impressed on each side of a groove. The two sides are cut 45° from the surface of the record.

forty-five record—A 7-inch record with a 1½-inch center hole. It is recorded at 45 rpm and played at the same speed.

four-channel (quadraphonic) sound—A system of sound reproduction using four separate signal paths from microphones to speakers (discrete system) or one of several matrixing systems to carry the same information on only two channels.

four-channel tape head—A tape head having four gaps and pickup coils and capable of playing quadraphonic program material.

four-track recording—Also called quarter-track. On ¼-inch tape, the arrangement by which four different channels of sound may

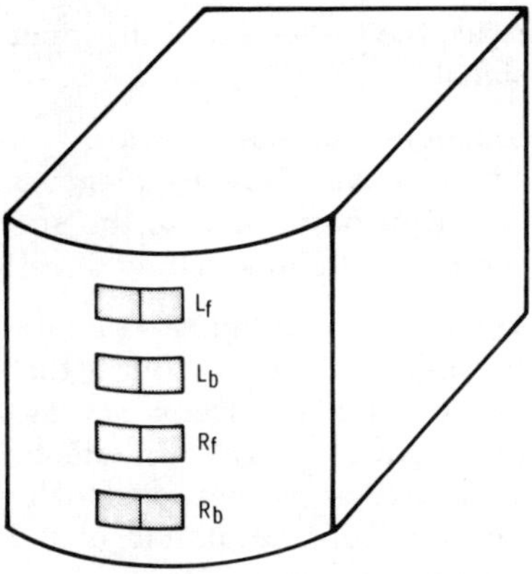

Four-channel tape head.

be recorded on adjacent tracks. These may be recorded as four separate and distinct tracks (monophonic) or two related pairs of tracks (stereo).

four-track tape—A tape with four recording tracks that permit it to be played in both directions for stereo recordings, like a two-track monophonic tape. Thus, twice as much material can be recorded as on a two-track stereo tape.

free impedance—Also called normal impedance. The input impedance of a transducer when the load impedance is zero.

free progressive wave—Also called free wave. A wave free from boundary effects in a medium. In other words, there are no reflections from nearby surfaces. A free wave can only be approximated in practice.

free reel—The reel that supplies the magnetic tape on a recorder.

free sound field—A field in a medium free from discontinuities or boundaries. In practice, it is a field in which the boundaries cause negligible effects over the region of interest.

free wave— *See* Free Progressive Wave.

freq—Abbreviation for frequency.

frequency—Symbolized by f. The number of recurrences of a periodic phenomenon in a unit of time. The old unit of frequency was the cycle per second. The new unit combines these two terms in a single term, the hertz, which is equivalent to one cycle per second. Radio frequencies are normally expressed in kilohertz at and below 30,000 kilohertz and in megahertz above this frequency. Audio frequencies are considered to be the range from about 15 hertz to 20,000 hertz.

frequency band—A continuous and specific range of frequencies.

frequency bias—A constant frequency purposely added to the frequency of a signal.

frequency compensation—The use of circuit elements or components to modify the frequency response of a system as a means of restoring losses or achieving a certain desired response.

frequency conversion—The process of converting a signal to some other frequency by combining it with another frequency.

frequency converter—Also called frequency changer. A circuit, device, or machine that changes an alternating current from one frequency to another, with or without a change in voltage or number of phases.

frequency correction—Compensation for unequal transmission of various frequencies in a line. Usually attained by means of an attenuation equalizer.

frequency deviation—In frequency modulation, the peak difference (in hertz) between the instantaneous frequency of the modulated wave and its carrier frequency.

frequency discriminator—A circuit that converts a frequency-modulated signal into an audio signal.

frequency distortion—The distortion that results when all frequencies in a complex wave are not amplified or attenuated equally.

frequency divider—A device that delivers an output voltage that is an integral submultiple or proper fraction of the input frequency.

frequency doubler—An electronic stage having a resonant output circuit tuned to the second harmonic of the input frequency. The output signal will then have twice the frequency of the input signal.

frequency - modulated wave— A carrier wave whose frequency is varied in proportion to the amplitude of the modulating signal.

frequency modulation—Modulation of a sine-wave carrier so that its instantaneous frequency differs from the carrier frequency by an amount proportional to the instantaneous amplitude of the modulating wave. Combinations of phase and frequency modulation also are commonly referred to as frequency modulation. (*See also* Frequency-Modulated Wave.)

frequency-modulation broadcast band—The band of frequencies from 88 to 108 megahertz set aside by the FCC for fm broadcasting. The band is divided into 100 channels, each 200 kHz in width.

frequency multiplex—A technique for the transmission of two or more signals over a common path.

frequency multiplier—A device for delivering an output wave with a frequency that is a multiple of the input frequency (e.g., frequency doublers and triplers).

frequency record—A recording of various known frequencies at known amplitudes, used for testing purposes.

frequency response—1. A measure of how effectively a circuit or device transmits the different frequencies applied to it. 2. The portion of the frequency spectrum that can be sensed by a device within specified limits of amplitude error.

frequency-response characteristic—The amount by which the gain or loss of a device varies with the frequency.

frequency - response curve— A graphical representation of the way a circuit responds to different frequencies within its operating range.

frequency-response equalization—Also called equalization or corrective equalization. The effect of all frequency-discriminative means employed in a transmission system to obtain the desired overall frequency response.

frequency selectivity—The degree to which a transducer is capable of separating a desired signal from signals or interference at other frequencies.

frequency tripler—An amplifier whose output circuit is resonant to the third harmonic of the input signal. The output frequency is three times the input frequency.

frequency-wavelength relation—For any propagated wave (either sound or electromagnetic) the wavelength is equal to the velocity of propagation divided by the frequency. Expressed as a formula, this is:

$$\lambda = \frac{c}{f}$$

where,

λ is the wavelength in any chosen unit of length,

c is velocity of propagation in the same unit of length per second,

f is frequency in hertz.

For electromagnetic waves (radio waves), c is 300,000,000 meters per second.

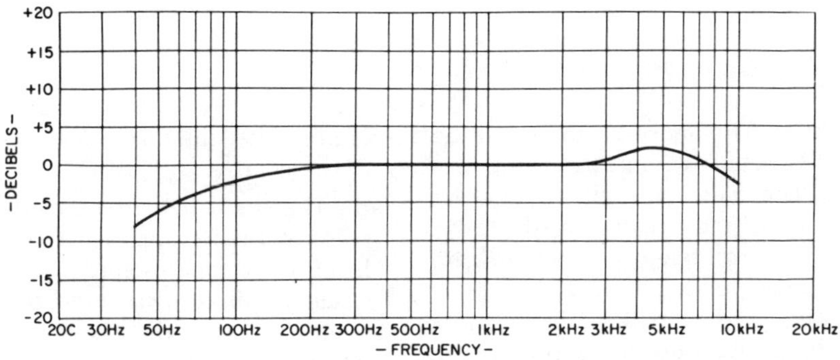

Frequency-response curve.

full-track recording—A magnetic recording in which the track width is essentially equal to the tape width. Applies to quarter-inch-wide (or less) tape only.

fundamental— *See* Fundamental Frequency.

fundamental component—The fundamental frequency component in the harmonic analysis of a wave.

fundamental frequency—The principal component of a wave; i.e., the component with the lowest frequency or greatest amplitude.

fundamental harmonic—The harmonic component with the lowest frequency.

fundamental mode—Of vibration, the mode having the lowest natural frequency.

fundamental tone—1. In a periodic wave, the component corresponding to the fundamental frequency. 2. In a complex tone, the component tone of lowest pitch.

fundamental wavelength—The wavelength corresponding to the fundamental frequency.

fuzz—An intentional distortion of the natural tone of an electric guitar. Also called fuzz tone.

G

gain—The increase in power between two points in an amplifier. The ratio of output to input. Usually expressed in dB. Losses are expressed as negative gain, or a gain less than 1.

gamma ferric oxide—The magnetic constituent of many present-day tapes, in the form of a dispersion of fine needlelike particles within the coating.

gap—The space between opposite poles of a recording or playback head in a tape recorder. The space is filled with some nonmagnetic material, and gap width is usually of the order of mils or microns.

gap depth—The dimension of the gap in a recording head, measured perpendicular to the surface of the head.

gap length—The dimension of the gap of a recording head, measured from one pole face to the other. In longitudinal recording, the gap length is the dimension of the gap in the direction of tape travel.

gap loss—The loss in output because of the finite gap length of the reproduce head. The loss increases as the wavelength decreases, amounting to approximately 4 dB when the wavelength is equal to twice the gap length and subsequently increases rapidly to a total loss of output when the wavelength is equal to approximately 1.15 times the gap length.

gap width—The dimension of the gap, measured in the direction parallel to the head surface and pole faces (perpendicular to the direction of tape travel).

gauss— The centimeter-gram-second electromagnetic unit of magnetic induction. One gauss represents one line of flux (one maxwell) per square centimeter.

gigahertz—One thousand million hertz or 10^9 hertz. Abbreviated GHz.

glissando—A succession of musical tones rising or falling in pitch and changing smoothly from one pitch to the next.

gradient microphone—A microphone in which the output rises and falls with the sound pressure. (*See also* Pressure Microphone.)

great manual—In an organ, the keyboard normally used for playing the accompaniment to the melody. Also called the accompaniment manual or lower manual.

groove—In mechanical recording, the track inscribed in the record by a cutting or embossing stylus, including any modulation caused by the vibration of the stylus. In stereo discs, its cross section is a right-angled triangle, with each side at a 45° angle to the surface of the record. Information is cut on both sides of the groove.

groove angle—In disc recording, the angle between the two walls of an unmodulated groove in a radial plane perpendicular to the surface of the recording medium.

groove shape—In disc recording, the contour of the groove in a radial plane perpendicular to the surface of the recording medium.

groove speed—In disc recording, the linear speed of the groove with respect to the stylus.

ground—1. A metallic connection with the earth to establish ground potential. 2. The voltage reference point in a circuit. There may or may not be an actual connection to earth, but it is understood that a point in the circuit said to be at ground potential could be connected to earth without changing the operation of the circuit in any way.

ground bus—A conductor, usually of large-diameter wire, that connects a number of points to one or more grounding electrodes.

grounded outlet—An outlet equipped with a receptacle having, in addition to the current-carrying contacts, one ground contact that can be used for the connection of an equipment-grounding conductor.

ground noise—In recording and reproducing, the residual noise in the absence of a signal. It is usually caused by dissimilarities between the recording and reproducing media, but may also include amplifier noise as from a tube or from a resistive element at the input of the reproducer amplifier system.

ground potential—Zero potential with respect to ground or the earth.

grouping—Nonuniform spacing between the grooves of a disc recording.

guard circle—An inner concentric groove on disc records. It prevents the pickup from possible damage by being thrown to the center of the record.

H

half cycle—One-half of a complete cycle, or 180° rotation of a generating vector, or 3.14 radians in radian measure.

half nut—In a disc-recording mechanism, a feed nut that engages half the circumference or less of a lead screw, so that it can be withdrawn from the lead screw to stop the lateral scanning movement.

half-power point—On an amplitude response curve or similar curve plotting power against frequency, angle, or other variable, the point that corresponds to half the power of a neighboring point having maximum power.

half step—*See* Semitone.

half-track recorder—*See* Dual-Track Recorder.

half-track tape—Also called two-track tape. Quarter-inch magnetic tape on which half the width of the tape is used for one sound path. Such a tape can provide stereo in one direction of tape travel, or mono sound in both directions.

hand receiver—An earphone held to the ear by hand.

handset—A telephone-type receiver and transmitter mounted on a single frame.

handset telephone—*See* Hand Telephone Set.

hand telephone set—Also called a handset telephone. A telephone set consisting of a handset and a mounting that supports the handset when not in use.

half wave—A wave with an electrical length of half a wavelength.

hangover—The tendency of the moving element of a speaker to continue moving after the applied signal drops to zero. This causes a blurred quality of the sound reproduction; individual tones tend to run together.

hard magnetic materials—Magnetic materials that are not easily demagnetized.

harmonic—A sinusoidal wave having a frequency that is an integral multiple of the fundamental frequency, such as 2, 3, or 4 times, etc. For example, a wave with twice the frequency of the fundamental is called the second harmonic.

harmonic analysis—1. A method of identifying and evaluating the harmonics that make up a complex waveform of voltage, current, or some other varying quantity. 2. The expression of a given function as a Fourier series of sine and cosine terms that are approximately equal to the given function.

harmonic analyzer—Also called harmonic-wave analyzer. A mechanical or electronic device for measuring the amplitude and phase of the various harmonic components of a wave.

harmonic component—Of a periodic quantity, any of the simple sinusoidal quantities of the Fourier series into which the periodic quantity may be resolved.

harmonic content—1. The degree of distortion in the output signal of an amplifier. 2. The components remaining after the fundamental frequency has been removed from a complex wave.

harmonic conversion transducer—A conversion transducer in which the useful output frequency is a multiple or submultiple of the input frequency.

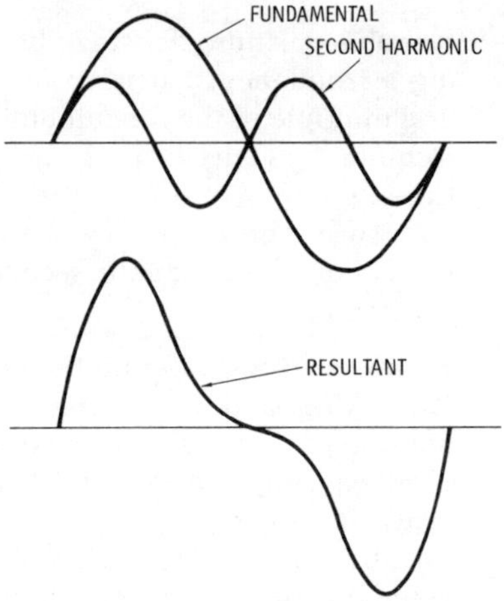

Harmonic content.

harmonic distortion—1. The production of harmonic frequencies in the output by the nonlinearity of a transducer when a sinusoidal voltage is applied to the input. The amplitude of the distortion is usually a function of the amplitude of the input signal. 2. A condition that exists in the output of an amplifying circuit when harmonics, added during the process, alter the signal waveform.

harmonic filter—A combination of inductance and capacitance tuned to an undesired harmonic to suppress it.

harmonic generator—A vacuum tube or other generator operated so that it generates strong harmonics in the output.

harmonic motion— Rectilinear motion to and fro past a point of equilibrium in such manner that acceleration is proportional to displacement from that point and is always directed toward the point. One example is the motion of a pendulum. The graph of simple harmonic motion plotted against time is a sine wave.

harmonic series of sounds—A series in which each basic frequency is an integral multiple of a fundamental frequency.

harmonic wave analyzer—*See* Harmonic Analyzer.

head—A device for reading, recording, or erasing information in a storage medium. Specifically, in tape recording, an electromagnet consisting of a ring-shaped metal core with a narrow gap of nonmagnetic material.

head alignment—Positioning the record-playback head on a tape recorder so that the head gap is perpendicular to the direction of tape travel.

head demagnetizer—A device, usually electromagnetic, for eliminating any residual magnetism built up in a recording head.

head, dual-track—A head with a pole piece that covers only half the recording width of the tape. The second track is recorded by flipping the tape and recording again.

head, full-track—A head with a gap that covers the full recording width of the tape.

head gap—A space inserted into the magnetic circuit of a magnetic-recorder head to direct the flux into the recording medium.

headphone—Also called a head receiver or phone. A sound transducer designed to be held against the ear and having a diaphragm that vibrates according to current variations in the audio signal applied to its coil windings.

head, quarter-track—A head with 2 pole pieces that cover the first and third quarters of the tape. The second and fourth quarters are recorded by flipping the tape and recording again.

head receiver—*See* Headphone.

headset—A headphone (or a pair of headphones) and its associated headband and connecting cord.

Headset (dynamic stereophonic headphones manufactured by Koss Electronics, Rek-O-Kut Div.).

head stack—A group of two or more heads mounted in a single unit, used to provide multiple-track recording or reproduction.

head-to-tape contact—The degree to which the surface of the magnetic coating of the tape touches the surface of the record or play heads of a recorder. Good head-to-tape contact minimizes separation loss and is essential for good high-frequency reproduction.

hearing loss—Lowered response of the ear at a specified frequency—i.e., the ratio in decibels of the threshold of audibility of the impaired ear to the threshold of the normal ear.

hearing loss for speech—The difference in decibels between the speech levels at which the average normal ear and the defective ear, respectively, reach the same intelligibility. It is often set at 50% (3-dB power loss).

heated stylus—In disc recording, a cutting stylus having provisions for heating the stylus during recording. In addition to normal cutting action of the stylus, the heat provides a partial melting of the disc lacquer in contact with the stylus and this, in turn, results in exceptionally smooth groove walls for noise-free reproduction.

hecto—A prefix meaning 100.

Helmholtz resonator—An acoustic enclosure with a small opening that causes the enclosure to resonate. The frequency of resonance depends on the area of the opening and the volume and shape of the resonator.

henry—The cgs electromagnetic unit of inductance or mutual inductance. The inductance of a circuit is 1 henry when a current variation of 1 ampere per second induces 1 volt. Abbreviated H.

hertz—Abbreviated Hz. A unit of frequency equal to one cycle per second.

heterodyne—Also called beat. To mix two frequencies together in a nonlinear device thereby producing other frequencies, including the sum and difference of the two original frequencies.

heterodyne frequency—The sum or difference produced by combining two other frequencies in a nonlinear device.

hf—Abbreviation for high frequency.

hi-fi—*See* High Fidelity.

high boost—*See* High-Frequency Compensation.

high fidelity—Popularly called hi-fi. The characteristic that enables an audio system to reproduce sound as nearly like the original as possible.

high filter—An audio circuit designed to remove undesired high-frequency noise such as record scratch, tape hiss, etc., from program material.

high-frequency bias—In a tape recorder, a sinusoidal voltage that is mixed with the signal being recorded to improve the linearity and dynamic range of the recorded signal. In practice, the bias frequency is three to four times the highest information frequency to be recorded.

high-frequency compensation—Also called high boost. An increase in the amplification of high audio frequencies, either to make up losses at certain points in a system or to change record or playback characteristics to gain certain advantages.

high-frequency unit—*See* Tweeter.

high-pass filter—A wave filter that passes all frequencies above a certain critical or cutoff frequency and attenuates those below that frequency, usually at a specified attenuation rate per octave.

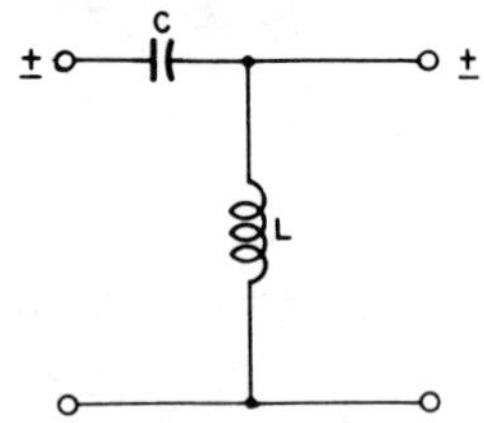

Configuration.

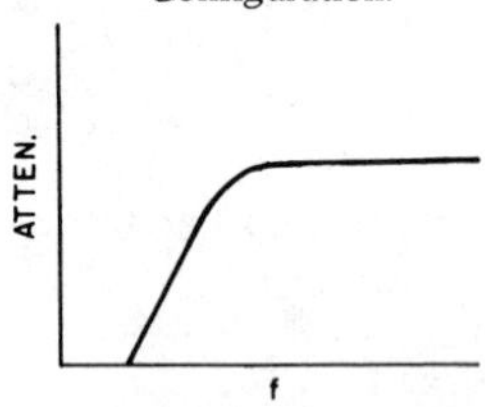

Transmission Characteristics.

High-pass filter.

high Q—Having a high ratio of reactance to effective resistance. The factor determining the efficiency of a reactive component.

hill-and-dale recording—*See* Vertical Recording.

hiss—Random noise characterized by prolonged sibilant sounds in the higher audio frequencies.

horn—Also called an acoustic horn. A tubular or rectangular enclosure for radiating or receiving acoustic waves.

horn cutoff frequency—A frequency below which an exponential horn will not function efficiently, because it fails to provide for proper expansion of the sound waves.

horn loading—A method of coupling the moving element in a speaker to the listening space by an expanding air column having a small throat and large mouth.

horn mouth—The wide end of a horn.

horn speaker—A speaker in which a horn couples the radiating element to the air space.

horn throat—The narrow end of a horn.

hot-wire microphone—A microphone in which the cooling or heating effect of a sound wave changes the resistance of a hot wire and thus controls the current through it.

hot-wire transducer—A transducer in which the heating or cooling effect of a sound wave changes the resistance of a hot wire and thus the current through it.

howl—An undesirable prolonged wail produced in a speaker by electric or acoustic feedback.

hum—Low-frequency background noise usually derived in some way from the alternating current supply line, with strong frequency components at 60 or 120 Hz.

hum-balancing pot—A potentiometer usually placed across the heater supply in vacuum tube circuits. Its arm is grounded so that the heater voltage may be balanced with respect to ground.

hum bucking—The introduction of a small amount of voltage, at the power-line frequency, into a circuit to cancel unwanted signal pickups from the power line.

hum-bucking coil—A coil wound around the field coil of a dynamic speaker and connected in series opposition with the voice coil. In this manner, any hum voltage induced in the field coil will be induced in the voice coil in the opposite direction and buck, or cancel, the effects of the hum.

hybrid—An electronic circuit that contains both vacuum tubes and transistors.

hydroacoustics—The generation of acoustic energy from the flow of fluids under pressure.

hydroacoustic transducer— A transducer that produces high-level acoustic energy from the flow of high-pressure fluid.

hydrophone—An electroacoustic transducer that responds to sound waves in water and delivers essentially equivalent electric waves.

hyperacoustic zone—The region in the upper atmosphere, above about 60 miles, in which the distance between air molecules is roughly equal to the wavelength of sound, so that sound is transmitted with less efficiency than at lower levels. Sound waves cannot be propagated above this zone.

hyperbolic grind—A term describing the contact area of certain tape record and playback heads. It permits good head-to-tape contact and good high-frequency response.

hyperbolic horn—A horn in which the equivalent cross-sectional radius increases according to a hyperbolic law.

hypersonic—Having five or more times the speed of sound.

hysteresigraph—A device for experimentally presenting or recording the hysteresis loop of a magnetic specimen.

hysteresis—1. The amount the magnetization of a ferrous material lags the magnetizing force because of molecular friction. 2. The difference between the response of a unit or system to an increasing or decreasing signal. 3. The lag in the response of an instrument or process when a force acting on it changes abruptly.

hysteresis curve—A curve showing the relation between a magnetizing force and the resultant magnetic flux.

hysteresis distortion—Distortion of waveforms in circuits containing magnetic components. It is caused by the hysteresis of the magnetic cores.

hysteresis error—The difference in the reading obtained with a measuring instrument containing iron when the current is increased to a definite value and the reading obtained when the current is reduced from a higher value to the same definite value.

hysteresis loop—For a magnetic material, a curve showing two values of magnetic flux density—

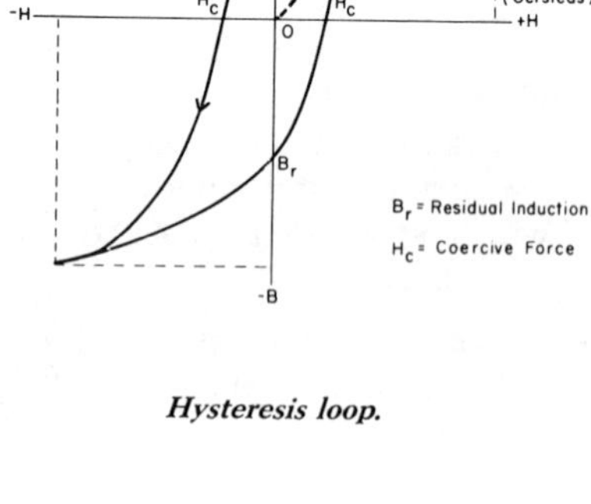

Hysteresis loop.

one when the magnetizing force is increasing and the other when it is decreasing—for each value of the magnetizing force.

hysteresis loss—The power expended in a magnetic material as a result of magnetic hysteresis.

hysteresis motor—Also called hysteresis synchronous motor. A synchronous motor without salient poles or direct-current excitation. It maintains its speed in step with the frequency of the current applied to operate it; hence, it is often used, with power-line current applied, as a constant-speed drive for tape recorders and record players.

Hz—Abbreviation for hertz.

I

I—Symbol for current.

idler—1. A rubber-rimmed wheel that transfers power from a phonograph motor to the turntable rim. 2. In tape recorders or record players, an intermediary wheel between drive motor and final driven device, and not directly connected to the power source.

IHFM—Abbreviation for Institute of High Fidelity Manufacturers, an association of manufacturers that publishes ratings and standards for high-fidelity equipment.

impedance—The total opposition (i.e., resistance and reactance) a circuit offers to the flow of alternating current at a given frequency. The ratio of the potential difference across a circuit or element of a circuit to the current through the circuit or element. It is measured in ohms, and its reciprocal is called admittance. Symbol: Z.

impedance angle—Angle of the impedance vector with respect to the resistance vector. Represents the phase angle between voltage and current.

impedance characteristic—A graph of impedance versus frequency of a circuit or component.

impedance coupling—A method of coupling using an impedance as the coupling device, common to both the primary and secondary circuits. This type of coupling is usually limited to audio systems, where high gain and limited bandpass are required.

impedance matching—The connection across a source impedance of another impedance having the same magnitude and phase angle. The object is maximum power transfer and fidelity of reproduction.

impedance-matching transformer—A transformer used to match the impedance of a source and load.

impedance triangle—A diagram consisting of a right triangle. The sides are proportional to the resistance and reactance in an ac circuit, with the hypotenuse representing the impedance.

index counter—An odometer-type counter that indicates the amount of tape that has been unreeled on a recorder. By its use, particular selections on a recorded tape can be indexed and returned to.

indicator tube—An electron-beam tube that visually conveys useful information (signal level, etc.) by the variation in cross section of the beam at a luminescent target.

indirect piezoelectricity—The production of a mechanical strain in a crystal by applying a voltage to it (as opposed to the more common practice of applying strain to a crystal in order to produce voltage).

inductance—In a circuit, the property that opposes change in the existing current. Inductance is present only when the current is changing.

inductance coil—*See* Inductor.

induction—1. The establishment of an electric charge or a magnetic field in a substance by the proximity of an electrified source, a magnet, or a magnetic field. 2. The generation of an electromotive force and current in a conductor by variation of the magnetic field affecting the conductor.

induction motor—An alternating-current motor in which the primary winding (usually the stator) is connected to the power source and induces a current into a polyphase secondary or squirrel-cage secondary (usually the rotor).

induction noise—The noise—other than thump, flutter, cross fire or cross talk—produced when two circuits are inductively coupled together.

induction speaker—A speaker in which the current that reacts with the steady magnetic field is induced into the moving member.

inductive—Pertaining to inductance or to the inducing of a voltage through mutual or electrostatic induction.

inductive circuit—A circuit with more inductive than capacitive reactance.

inductive coupling—The use of one circuit to influence a similar circuit by means of inductance common to both.

inductive feedback—The transfer of energy from the output circuit to the input circuit of an amplifying device through an inductor or inductive coupling.

inductive load—Also called lagging load. A load that is predominately inductive, so that the alternating load current lags behind the alternating voltage of the load.

inductively coupled circuit—Two adjacent parts of a circuit coupled by mutual inductance.

inductive microphone—*See* Inductor Microphone.

inductive pickup—An action whereby signals are generated in a circuit or conductor because of mutual inductance between it and a disturbing source.

inductive reactance—The opposition to the flow of alternating or pulsating current by the inductance of a circuit. It is measured in ohms, and its symbol is X_L. Its value is found by $X_L = 2\pi fL$, where f is frequency in hertz and L is inductance in henrys.

inductive transducer—A transducer in which a change in inductance is the link between input and output.

inductor—Also called inductance or retardation coil. A conductor used for introducing inductance into an electric circuit. The conductor is wound into a spiral, or coil, to increase its inductive strength.

inductor microphone—Also called inductive microphone. A microphone in which the vibration of sound waves causes a conductor to move back and forth in a magnetic field, creating an electrical output.

inertance—Acoustical equivalent of inductance.

inertia—The tendency of an object at rest to remain at rest, or of a moving object to continue moving in the same direction and at the same speed, unless disturbed by an outside force.

infinite baffle—A speaker enclosure with no openings through which sound waves can travel from the front of the speaker cone to the back.

infrasonic—Pertaining to frequencies below the range of human hearing, or below about 15 hertz. Formerly called subsonic.

infrasonic frequency—A frequency below the normal range of human hearing.

inharmonic frequency—A frequency that is not a rational multiple of another frequency.

in-line heads— *See* Stacked Heads.

in phase—Two waves of the same frequency that pass through their maximum and minimum values of like polarity at the same instant are said to be in phase.

input device—The device or set of devices through which an input signal is applied to another device.

input impedance—1. The impedance a transducer presents to a source. 2. The effective impedance "seen looking into" the input terminals of an amplifier.

inside spider—A flexible device placed inside a voice coil to center it with the pole pieces of a speaker.

input transformer—A transformer that transfers energy from an alternating-voltage source to the input of a circuit or device. It is usually chosen to provide the correct impedance match, also.

insertion loss—The difference between the power received at the load before and after the insertion of apparatus at some point in the line.

instantaneous companding—Companding that varies the effective gain in response to instantaneous values of the signal wave.

instantaneous disc—A blank recording disc that can be played back on a phonograph immediately after being cut on a recorder.

instantaneous power output—The rate at which energy is delivered to a load at a particular instant.

instantaneous recording—A recording intended for direct reproduction without further processing.

instantaneous sound pressure—The total instantaneous pressure at a certain point, minus the static pressure at that point. Commonly measured in microbars.

instantaneous speech power—The rate at which a speaker is radiating sound energy at any given instant.

instantaneous value—The magnitude, at any particular instant, of a varying value.

integer—A whole number; that is, not a simple fraction or a mixed fraction.

integrated amplifier—An audio amplifier unit made up of both a preamplifier and a power amplifier on one chassis.

integrated circuit—Abbreviated IC. A combination of interconnected circuit elements manufactured as a unit on a continuous substrate.

integrated transducers—Standard types of semiconductor transducer (thermistors, photogenerators, piezoelectric devices, etc.)

incorporated during manufacture in integrated multifunction chips.

intelligence signal—Any signal that conveys information (e.g., voice, music, code, or television).

intensity level—Ratio, under normal conditions of the intensity of a sound to a reference intensity of a free plane wave of 1 microwatt per square centimeter. Commonly expressed in decibels.

interaction—The effects two or more parts, components, etc., have on each other while each is performing a function.

intercom—*See* Intercommunication System.

intercommunication system—Also called intercom. A two-way communication system operating without a central switchboard. Often consists of a master station and several slave stations. Slave stations may or may not be equipped to originate calls, but can answer calls from the master station.

interelectrode capacitance—The capacitance between one electron-tube electrode and an adjacent electrode.

interelectrode coupling—Transfer of signal between elements of an electron tube by virtue of interelectrode capacitance. If signal transfer is from output to input and of positive phase, oscillations may develop, thus limiting amplification.

interelectrode leakage—The undesired current that flows between elements not normally connected in any way.

interelectrode transit time—The time required for an electron to travel between two electrodes.

interelement capacitance—The capacitance caused by the pn junctions between the regions of a transistor and measured between the external leads of the transistor.

intermodulation—In a nonlinear signal-handling device, the production of frequencies corresponding to the sums and differences of two or more frequencies transmitted through the device.

intermodulation distortion—The introduction of new, undesired frequencies in the output of a device, caused by passing two or more frequencies together through nonlinear portions of the device.

intermodulation frequencies—The sum and difference frequencies generated in a nonlinear element.

internal input impedance—The actual impedance at the input terminals of a device. Can be represented by the ratio of input voltage to input current that results.

internal output impedance—The

actual impedance at the output terminals of a device.

internal resistance—Any built-in characteristic of a voltage source that opposes unlimited delivery of current by the source.

interrecord gap—Also called inter-block space. The space between recorded sections on magnetic tape caused by delays in starting and stopping the tape motion. This gap signals that the end of a record has been reached.

interstage transformer—A transformer used between stages to couple two stages together.

intrinsic coercive force—The magnetizing force required to reduce the intrinsic induction of a magnetized material to zero, when applied in the opposite direction.

intrinsic induction—The excess magnetic induction produced in a magnetic material by a given magnetizing force, as compared to the induction that would be produced by the same magnetizing force in a vacuum.

intrinsic noise—Noise that is caused by a device itself or by the transmission path, independent of modulation.

intrinsic permeability—The intrinsic normal induction divided by the corresponding magnetizing force.

inverse feedback—*See* Negative Feedback.

inverse-feedback filter—A tuned circuit in the feedback line of a highly selective amplifier having negative feedback. The feedback output is zero for the resonant frequency, but increases rapidly on either side of this frequency.

inverse piezoelectric effect—Motion in a piezoelectric crystal (contraction or expansion) under the influence of an electric field.

inverter—1. A circuit that produces a positive output from a negative input, or vice versa. 2. A device that changes alternating current to direct current or vice versa.

inverting amplifier—An amplifier in which the output polarity is the opposite or inverse of the input polarity.

ion—An atom that has either gained or lost electrons. Positively charged ions have a deficiency of electrons and negatively charged ions have a surplus of electrons.

ionic tweeter—A type of speaker whose moving element is a mass of ionized air, activated by a varying electrostatic field. Because the ionic "diaphragm" is very light in weight, the speaker is capable of extremely extended high-frequency response (up to 100 kHz or so).

ionization—The dissociation of an atom or molecule into electrons or ions.

ionize—To remove electrons from an atom or molecule (e.g., by Xray bombardment) thus transforming the atom or molecule into a positive ion. The freed electron attaches itself to another atom or molecule, which thus becomes a negative ion.

ionophone— A high-frequency speaker in which the moving element is a modulated arc in the mouth of a quartz tube. The resultant modulated wave acts directly on the ionized air under pressure and thus creates sound waves.

ips—Abbreviation for inches per second.

IR drop—The voltage or drop in potential across a resistance (R) when there is a current (I) through the resistor.

iron-core coil—A coil in which a core of soft iron is used to intensify the magnetic field created by the coil winding. The iron core may be in the form of solid iron, powdered iron, laminated sheets, etc.

iron-core transformer—A transformer in which iron forms part or all of the magnetic circuit linking the transformer windings.

iron loss—*See* Core Loss.

iron oxide—One of the materials used as a recording base on recording tape.

isochronous circuits—Literally, having equal time. Said of two circuits having the same resonant frequency.

isolation—Electrical or acoustical separation between two locations.

isolation amplifier—An amplifier employed to minimize the effects of one circuit on another.

isolation transformer—A transformer designed to provide ac coupling between two or more circuits (for example, an amplifier and the power line) while at the same time offering protection against overloads, shock hazards, and other disturbing effects.

isotropic—Having properties with the same values in all directions along axes.

isotropic magnet— A magnetic material having magnetic properties that are the same in all directions.

I^2R—Power in watts expressed in terms of current (I) and resistance (R).

I^2R loss—The power lost in transformers, generators, connecting wires, and other parts of a circuit because of current through the resistance of the conductors.

J

J—Abbreviation for joule.

jack—A socket that can be wired into a circuit, with provision to accept a plug, thus allowing quick connect or disconnect to the circuit.

jack panel—An assembly made up of a number of jacks mounted on a board or panel.

JFET—Abbreviation for junction field-effect transistor.

Jones plug—A polarized connector having the form of a receptacle with a number of contacts.

joule—1. The work done by a force of 1 newton acting through a distance of 1 meter. The joule is the unit of work in the mksa system. 2. The energy required to transport a charge of 1 coulomb between two points having a potential difference of 1 volt. The joule is 10^7 ergs. A kilowatt-hour is 3.6×10^6 joules.

junction field-effect transistor—A transistor with drain, source, and gate terminals. The gate is insulated from drain and source and signal is applied to it. The electric field created by the gate influences the current through drain and source. Compared to the common junction transistor, the field-effect transistor has a high input impedance. (Abbreviated JFET.)

junction transistor—A transistor having two parallel pn junctions created by either pnp or npn sections of semiconductor material.

just scale—A musical scale formed by using intervals between tones whose frequencies have integral ratios, such as 3/2, 4/3, 5/4, 5/3, 9/8, etc.

K

keeper—A bar of easily magnetized material (usually soft iron) placed across the poles of a permanent magnet to protect it from demagnetization.

kHz—Abbreviation for kilohertz.

kilo—A prefix representing 10^3, or 1000. Abbreviated k.

kilohertz—1000 hertz.

kilohm—1000 ohms. Abbreviated K or kΩ.

kilovolt—1000 volts. Abbreviated kV.

kilowatt—1000 watts. Abbreviated kW.

kilowatt hour—A unit used to measure electrical energy. The equivalent of energy supplied at a rate of 1000 watts for a period of one hour.

kinetic energy—Energy stored in a moving system because of its motion. The energy is released as the system comes to rest.

L

L—Symbol for coil or inductance.

labyrinth—Speaker enclosure embodying a long, folded pathway to delay the rear wave from the speaker unit.

lacquer disc—Also called cellulose-nitrate disc. A disc for mechanical recording, usually made of metal, glass, or paper and coated with a lacquer compound often containing cellulose nitrate.

lacquer master— *See* Lacquer Original.

lacquer original—Also called lacquer master. An original recording cut on a lacquer recording blank for use as a master.

lacquer recording—Any recording made on a lacquer medium.

lag—The time displacement by which one wave follows another wave of the same frequency. Usually stated in electrical degrees.

lagging current—Current that passes through corresponding maximum and minimum values at a later time than the maximum and minimum values of the applied voltage causing the current. Lag is usually stated in electrical degrees. Lagging current is characteristic of a circuit that is predominately inductive.

lagging load—A predominately inductive load—i.e., one in which the current lags the voltage.

lambda—Greek letter λ, used as a symbol for wavelength.

laminated—Constructed of a number of layers.

laminated core—A core made of a number of thin sheets insulated from each other. A laminated iron core is used in transformers, coils, armatures, etc., to reduce eddy current losses.

laminated record—A record made of several layers of material (normally a core of inexpensive material sandwiched between two layers of higher quality).

lamination—A single stamping of sheet material to be used with similar stampings to build a laminated object such as the core of a power transformer.

land—1. The flat surface between two adjacent grooves on a recording disc. 2. Also called boss, blivet, donut, pad, spot, tab, and terminal point. The conductive area in a printed-circuit board, to which components or separate circuits are attached.

lapel microphone—A microphone worn on the user's clothing, often on the lapel.

large-signal characteristics—Amplifier characteristics obtained when input signal is great enough to produce full rated output.

large-signal power gain—The ratio of the full rated output power to the input power necessary to produce it. Usually expressed in dB.

laryngaphone—Also called a throat microphone. A microphone worn in contact with the throat of a speaker to pick up voice vibrations directly. Useful in reducing pickup of background noises.

lateral compliance—The ease with which the reproducing stylus can be moved from side to side as it follows the modulation on a laterally recorded record.

lateral recording—A mechanical recording in which the groove modulation is perpendicular to the direction of groove travel and parallel to the surface of the record.

lavalier microphone—A small microphone suspended from the neck by a cord or wire. Useful in broadcasting or public address, where the speaker needs to be free to move about, with free use of his hands.

law of electric charges—Like charges repel; unlike charges attract. (*See also* Coulomb's Law.)

law of electromagnetic induction—*See* Faraday's Law.

law of electromagnetic systems—Every electromagnetic system tends to change its configuration so that magnetic flux in the system will be a maximum.

law of electrostatic attraction—*See* Coulomb's Law.

law of magnetism—Like poles repel; unlike poles attract.

layer-to-layer adhesion—The tendency for adjacent layers of a roll of recording tape to stick together. This condition is more common after prolonged storage under conditions of high temperature or humidity.

layer-to-layer signal transfer—Signal transfer by the magnetic field recorded on one layer of tape affecting the layer next to it on the reel. Transfer tends to be greater if the recorded reel is stored for extended periods of time at high temperatures.

LC product—Inductance (L) in henrys multiplied by capacitance (C) in farads.

L/C ratio—Inductance (L) in henrys divided by capacitance (C) in farads.

LCS—Abbreviation for loudness-contour selector.

leader—The blank section at the beginning of a tape.

leader tape—Also called timing tape. Plain nonmagnetic tape for splicing to either end of magnetic tape to save the recorded tape from the wear and tear of repeated threading operations. Can also be useful as markers at selected points in the recorded material, or to insert silent spaces.

lead-in groove—Also called a lead-in spiral. A blank spiral groove at the beginning of the record. It provides a set-down point for the pickup and a quick lead-in to the recorded portion of the record.

lead-in spiral—*See* Lead-In Groove.

lead-over groove—Also called a crossover spiral. On disc records containing several selections, the groove in which the needle travels as it crosses from one selection to the next.

lead screw—In disc recording, a threaded rod that controls the rate of cutter travel across the recording disc.

leakage flux—The flux that does not pass through the air gap or the intended path of the magnetic circuit.

left (or right) signal—The electrical output of a microphone (or combination of microphones) placed so as to pick up sounds predominately to the listener's left (or right).

left (or right) stereophonic channel—The section of a stereophonic amplifier that receives the signal recorded from the left (or right) microphone of a stereo recording system and applies the amplified signal to the left (or right) speaker in the playback system.

Lenz's law—Lenz's law states that the current induced in a circuit because of a change in the magnetic flux through it or because of its motion in a magnetic field is in a direction that opposes the change in flux or that exerts a mechanical force opposing the motion.

level—The value of a quantity as compared to an arbitrary reference value. Level may be expressed in the same units as the quantity being measured (e.g., volts, ohms, etc.) or in dB compared to a specific reference level.

level above threshold—Also called sensation level. The level of a sound in dB above the threshold of audibility for the individual listener.

level indicator—1. A device on a tape recorder that indicates the level of the signal being recorded, thus serving as a warning against over- or under-recording. 2. A meter or other indicating device that shows the magnitude of signal at the point of application.

light-beam pickup—A phonograph pickup that uses a beam of lght as a coupling element between the record and the pickup.

limiter—A device used to automatically prevent the output from an applied signal from rising above a certain value, regardless of increase in signal amplitude.

limiting—The restriction of the amplitude of a signal so that the output does not rise above a certain value.

limit switch—A mechanically operated switch mounted in the path of a moving mechanism and designed to operate when a certain point of travel is reached. Switch operation can be chosen to either open or close a circuit. Example: The switch that automatically shuts off some record players at the end of a record.

line amplifier—An amplifier designed to match the impedance of a transmission line or system and supply a signal of specified level.

linear—Having an output that varies in direct proportion to the input. A graph of output versus input is a straight line.

linear acceleration—The rate of change in linear velocity.

linear amplification—Amplification in which the output is directly proportional to the input.

linear amplifier—An amplifier that amplifies all frequencies equally within the range of its response.

linear circuit—A circuit whose output is directly proportional to the input voltage.

linear control—A rheostat or potentiometer having a resistance uniformly distributed over the range of control-shaft rotation.

linear differential transformer—An electromechanical transducer that converts motion into output voltage, with phase and amplitude proportional to position. (*See also* Linear Motion Transducer.)

linearity—A relation between two quantities such that a change in one quantity produces a directly proportional change in the second quantity. If one is plotted against the other in a graph, a straight line will result.

linear motion transducer—A transducer that converts straight-line (linear) mechanical motion into an electrical ac analog signal.

linear power amplifier—A power amplifier in which the output voltage varies in direct proportion to the input voltage.

linear taper—The resistance characteristic of a potentiometer that changes resistance linearly as the control shaft is rotated through its range.

linear transducer—A transducer having an output that varies linearly with respect to the input.

linear velocity transducer—A transducer that produces an output signal proportional to the velocity of its moving element.

line cord—Also called a power cord. A two- or three-wire cord for connecting a radio, tv, amplifier, record player, etc., to the electric service line by plugging into a wall socket or similar receptacle.

line filter—A device containing a network of inductors and capacitors, designed for insertion between a transmitter, receiver, or electrical appliance and the power line to block noise signals, either from the appliance or from the power line.

line frequency—The frequency of the power-line voltage.

line microphone—A directional microphone made up of a single straight-line element and adjacent electroacoustic transducing elements in a straight line, or the acoustical equivalent of such an array.

line of force—An imaginary line in an electric or magnetic field and having the same direction as the field intensity at every point. When it is used as a unit of magnetic flux, it is sometimes called a maxwell.

line voltage—The voltage supplied to the consumer by the power company.

linkage—1. An indication of the voltage that will be induced in a circuit by magnetic flux. It equals the flux times the number of turns linked by the flux. 2. A mechanical arrangement of solid pieces with movable joints for transferring motion from one mechanism to another.

lip microphone—A sensitive microphone that is placed in contact with the lips.

liquid rheostat—A rheostat consisting of metal plates immersed in a conductive liquid. The resistance is changed by raising or lowering the plates in the liquid or by changing the distance between them (they must not be allowed to touch each other, however).

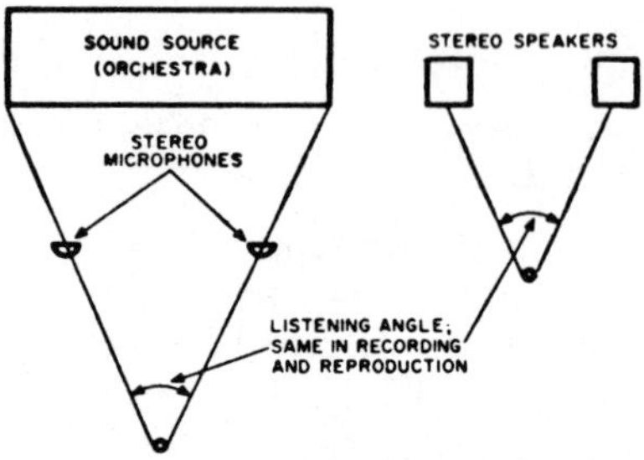

Listening angle.

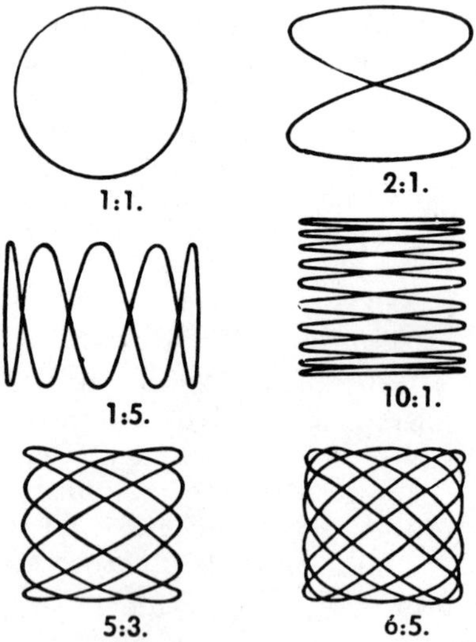

Lissajous figures.

Lissajous figures— Oscilloscope patterns produced when sine-wave signal voltages of various amplitude and phase relationships are applied to the horizontal- and vertical-deflection circuits simultaneously.

listening angle—The enclosed angle between the listener and the two speakers of a stereo reproducing system.

live room—A room in which sound-absorptive material such as drapery, rugs, upholstered furniture, etc., has been kept to a minimum. This allows reflective surfaces to predominate, and sounds will fade slowly, giving the room a live quality.

load—1. The power used by a machine or circuit in operation. 2. The power delivered by a machine. 3. A device that absorbs power and converts it into the desired form. 4. The impedance to which energy is being supplied.

loaded impedance—The impedance presented by the input of a transducer when the output is connected to its normal load.

loaded motional impedance—*See* Motional Impedance.

load impedance—The impedance that a load presents to a source from which it is receiving energy. This impedance and the internal impedance of the source determine the amount of energy transferred.

loading—Placing some material at the front or rear of a speaker so as to change its acoustic impedance and thus alter its radiation pattern.

loading coil—An inductor inserted into a circuit to increase its inductance and thereby improve its transmission characteristics.

load line—1. A straight line drawn through the operating point on a series of plate-current/ plate-voltage characteristic curves to show how plate current changes with grid voltage for a given plate load. 2. A line drawn on the collector characteristic curves of a transistor on which the operating point of the transistor moves as the collector current changes.

load-matching network—An electrical impedance network used when a source and its load are of different impedances. The matching network inserted between source and load provides for maximum transfer of energy.

locked groove—Also called concentric groove. The blank and continuous groove in the center, near the label, of a disc record. It prevents the stylus from traveling farther inward, onto the label.

logarithmic amplifier—An amplifier with an output that is a logarithmic function of the input.

logarithmic curve—A curve plotted on XY coordinates where all Y values of the curve are logarithms of the X value.

logarithmic scale—A scale on which the divisions are plotted according to the logarithm of the numbers with which the divisions are labeled. Example: the common slide rule.

logic circuit—A circuit used in quadraphonic sound systems to improve the channel separation in matrix-encoded material. The basic design feature is a variable-gain amplifier that detects which channel signal is dominant at the moment and automatically adjusts to favor that channel.

log/linear preamplifier—A preamplifier having an electrical output proportional to the common logarithm of the input, but with provision for switching to operation with output proportional to the input.

longitudinal magnetization—In magnetic recording, magnetization of the recording medium in a direction parallel to the line of travel.

longitudinal wave—A wave in which the direction of displacement of the particles of the elastic medium (usually air) is parallel to the direction of propagation of the wave.

long-play record—Abbreviated lp record. A name applied to the successors to the 78-rpm record. A combination of slower rotating speeds (45 rpm, 33 $^{1}/_{3}$, etc.) with finer and more closely spaced grooves results in greatly increased playing time. Also called a microgroove record.

long-pull magnet—An electromagnet with a physical construction that allows considerable armature travel before it closes the air gap. The armature is a conical plunger that moves in a core having a hollow space to fit the armature shape.

long throw—A method of speaker design using long voice coils and flexible suspension that allows long voice-coil excursions. This permits transfer of comparatively large amounts of acoustic power at low audio frequencies.

loop gain—The increase in gain observed when the negative feedback in an amplifier is disconnected, while all circuit loads remain intact.

loss—A decrease in power between two points in a circuit or amplifier. It is usually expressed in decibels.

loudness—A psychological property of sound whereby the listener judges the relative intensity of a sound. It is dependent mainly on sound pressure but is also influenced to some extent by frequency and complexity of waveform. The unit of loudness is the sone. The sone is the loudness of a 1000-Hz tone that has an intensity of 40 dB above the listener's own threshold of hearing.

loudness contour—A curve plotting sound pressures required at each frequency to produce a given loudness sensation to a typical listener.

loudness-contour selector—Abbreviated LCS. A circuit for changing the frequency response of an amplifier so that the amplifier response characteristics will more closely match the requirements of the human ear.

loudness control—Also called compensated volume control. A combined volume and tone control that boosts the bass frequencies at low volume to compensate for the lowered sensitivity of the ear at these frequencies. Some loudness controls provide similar compensation at the treble frequencies.

loudness level—The loudness level of a particular sound is equal to the sound-pressure level of a

1000-Hz tone judged by a listener to be just as loud. It is measured in decibels relative to 0.0002 microbar.

loudspeaker—*See* Speaker.

loudspeaker-dividing network—*See* Dividing Network.

loudspeaker impedance— *See* Speaker Impedance.

loudspeaker system—*See* Speaker System.

loudspeaker voice coil—*See* Speaker Voice Coil.

louver—The grille of a speaker.

lower manual—*See* Accompaniment Manual.

lower pitch limit—The lowest frequency at which a sinusoidal sound wave will produce a sensation of pitch.

lower sideband—The lower of two frequencies or groups of frequencies produced when the carrier frequency and the modulating frequency are combined in the amplitude-modulation process.

low filter—A filter circuit designed to remove low-frequency noises from the program material. Such noises include turntable rumble, tone-arm resonance, etc.

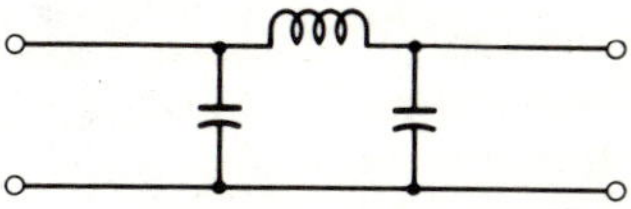

Low-pass filter.

low-frequency compensation—A circuit designed to improve the low-frequency response of a wideband amplifier.

low-pass filter—A filter network that passes all frequencies below a specified frequency (the cutoff frequency) but restricts the frequencies above that frequency.

low-print tape—Special magnetic recording tape with low print-through (the transfer of signal from one layer to another during storage).

L-pad—A volume control designed to present the same impedance throughout its control range. It consists essentially of an L-network in which both elements are adjusted simultaneously.

L+R, L–R—The sum and difference signals of the two stereo channels; left plus right and left minus right. The right or the left signal alone can be obtained by proper combination of these two signals; their sum (L+R+ L–R) produces 2L; their difference (L+R) − (L–R) produces 2R.

lp record—Abbreviation for long-playing record.

M

macrosonics—The use of high-amplitude sound waves for performing functions such as cleaning, drilling, emulsification, etc.

magamp—Abbreviation for magnetic amplifier.

magnet—A body having the property of attracting iron and similar magnetic materials and capable of producing a magnetic field. The natural form of magnet is called a lodestone.

magnetic—Pertaining to magnetism.

magnetic amplifier—A device in which one or more saturable reactors are used to obtain power gain. Often called magamp.

magnetic-armature speaker—Also called magnetic speaker. A speaker in which part of the moving element is a ferromagnetic material.

magnetic biasing—In tape recording, the linear mixing of a high-frequency magnetic field with the magnetic field of the signal to improve recording linearity.

magnetic cartridge—*See* Variable-Reluctance Pickup.

magnetic circuit—The path of magnetic flux from the north pole of the magnet, through the intervening materials of greatest permeability, to the south pole of the magnet, and internally through the magnet to its north pole. If the magnetic circuit is compared to an electrical circuit, magnetomotive force compares to voltage, flux lines compare to current, and reluctance compares to resistance.

magnetic coated disc—A magnetic disc-recording medium consisting of a coat of magnetizable material over a nonmagnetic base.

magnetic core—A geometrical structure of magnetic material designed to be placed in a magnetic circuit to direct or intensify the magnetic field.

magnetic cutter—A record cutter in which the recording stylus is deflected by the action of magnetic fields.

magnetic cycle—A complete series of changes in the magnetization of an object corresponding to a similar series of changes in one cycle of the alternating current producing the magnetization.

magnetic damping—The use of the reaction between a magnetic field and a conductor moving in that field to reduce or damp mechanical motion. An example is found in some meter movements where the moving coil form is made of a conductive material. Eddy currents in the coil form react with the magnetic field to reduce swing of the meter pointer.

magnetic density—The number of magnetic lines of force per unit area of cross section in a magnetic field.

magnetic disc—A disc used for magnetic recording. A flat circular disc is coated with material that can be easily magnetized for storing data to be played back later.

magnetic field—Any area where magnetic forces can be detected, i.e., the area around a permanent magnet, natural magnet, or electromagnet.

magnetic field strength—*See* Magnetizing Force.

magnetic flux—The magnetic induction in a material. In a magnetic circuit, this is the analog of electrical current in an electric circuit.

magnetic flux density—*See* Magnetic Induction.

magnetic gap—The nonmagnetic part of a magnetic circuit. The gap can be air or some other material commonly considered to be nonmagnetic.

magnetic head—In magnetic recording, a transducer for converting current variations into a varying magnetic field for storage on the magnetic recording medium or for reconverting previously stored fields into a varying current for amplification and playback.

magnetic hysteresis—In a magnetic material, the property of the material whereby the intensity of magnetization depends not only on the magnetizing force H, but also on the previous magnetic history of the material. That is, once a certain force H has changed the intensity of magnetization from zero to a particular value, it requires a greater force in the opposite direction to return the magnetic intensity to zero again.

magnetic hysteresis loop—A closed curve (commonly called a B-H curve) that shows the magnetic flux density in a magnetic substance as a function of magnetization force for a complete cycle of the magnetization force.

magnetic hysteresis loss—The power expended in a magnetic material, as a result of magnetic hysteresis, when the magnetic induction is cyclic.

magnetic induction—Also called magnetic flux density. The flux per unit area perpendicular to the direction of flux. The cgs unit of induction is called the gauss (plural, gausses) and is defined by the equation:

$$B = \frac{d\Phi}{dA}$$

magnetic instability—The tendency of a magnetic material to demagnetize or change in magnetic intensity with time, mechanical flexing, jarring, temperature changes, etc.

magnetic leakage—The straying of magnetic flux from the path designed for it in the magnetic circuit.

magnetic lines of force—Imaginary lines used to describe the strength and direction of a magnetic field.

magnetic materials—Materials capable of being magnetized, either by the influence of a permanent magnet or by the magnetic field developed by an electric current. Ferromagnetic materials show the strongest magnetic effects.

magnetic microphone—*See* Variable-Reluctance Microphone.

magnetic pickup—*See* Variable-Reluctance Pickup.

magnetic poles—Those regions of a magnet where the magnetic field appears to be most concentrated. In bar and horseshoe magnets, the poles are at the ends of the magnet.

magnetic-powder-coated tape—Also called coated tape. A tape made of some flexible, nonmagnetic base coated with a finely powdered, uniformly dispersed, ferromagnetic material.

magnetic recorder—A device for changing varying electrical signals to varying magnetic fields and impressing these fields on some type of magnetic recording medium for storage and eventual recovery of the electrical signals.

magnetic recording—Recording electrical signals by magnetizing areas of a tape or wire. The magnetized tape or wire is played back by passing it through a reproducing head, which recovers the original signals.

magnetic recording head—In magnetic recording, a device that transforms electric variations into magnetic variations for storage on a magnetic medium such as tape or discs.

magnetic recording medium—Magnetizable material in the form of tape, wire, a cylinder, a disc, or other appropriate form, for accepting and retaining magnetic variations impressed on it by the recording head.

magnetic reproducer—Equipment for scanning recorded magnetic variations on tape or other media and converting them into electrical variations.

magnetic reproducing head—In magnetic recording, the head that converts the magnetic variations into electrical variations.

magnetic saturation—In a magnetizing procedure, the point where further increase in magnetizing force produces little or no increase in magnetic lines of force.

magnetic shield—A device for shunting stray magnetic fields around instruments or radio parts that it is necessary to protect. The shield is usually a sheet or core of highly permeable material that provides an easy path for the magnetic lines of force.

magnetic speaker—A speaker that converts electrical impulses to magnetic impulses that in turn produce mechanical movement and sound waves.

magnetic tape—The recording medium used in tape recorders. It consists of a magnetic emulsion coated on a flexible plastic or paper base.

magnetic transducer—*See* Variable-Reluctance Transducer.

magnetic units—Units used in measuring magnetic quantities. Ampere-turn, gauss, gilbert, line of force, maxwell, oersted, and unit magnetic pole are examples of magnetic units.

magnetism—A property exhibited by both magnets and electric currents whereby certain materials such as iron are attracted or repelled and magnetic fields of force are developed.

magnetization curve—A plotted curve showing successive states during magnetization of a ferromagnetic material.

magnetizing force—Also called magnetic field strength. The magnetomotive force per unit length at any given point in a magnetic circuit. The magnetizing force in oersteds is found by the formula:

$$\frac{\text{Magnetomotive force in gilberts}}{\text{Length in centimeters}}$$

magnetomotive force—The force that produces a magnetic field. A magnetic field is produced either by an electric current through a conductor or by the proximity of a magnetized body. The cgs unit of magnetomotive force is the gilbert, equal to 0.7958 ampere turn.

magnetostriction—A property of certain materials that causes them to shrink or expand when placed in a magnetic field. Conversely, if subjected to compression or tensions, they experience a change in magnetic reluctance. Some materials that exhibit these effects are iron, nickel, cobalt, and some manganese alloys.

magnetostriction speaker— A speaker in which sound waves are generated through the deformation of a magnetostrictive material.

magnet steel—A special steel of high retentivity used in permanent magnets. In addition to steel, it contains tungsten, cobalt, chromium, and manganese.

manual—A group of organ keys played by the hand. The upper manual of two-manual organs is also referred to as the solo or swell manual and is normally used to play the melody. The lower manual, also called the accompaniment manual or great manual, is normally used to play the accompaniment.

masking— The phenomenon whereby one sound is obscured or "covered up" by another sound, called the masking sound. The degree of masking is usually measured in decibels.

mask microphone—A microphone designed for use inside an oxygen mask or other respiratory masks.

master—The original recording, which may be a disc, film, or magnetic tape. In disc recording, the next step is a metal matrix (negative; that is, grooves become ridges and ridges become grooves). The matrix is used to make the final vinyl pressings, if less than 200 pressings are to be made. If more than this number of pressings are planned, the matrix is used to make a "mother" (positive) from which "stampers" can be made, and from these the final records are pressed. In either case, the original master record is preserved undamaged.

matched impedance—The condition existing when two coupled circuits have equal impedances, or, if unequal, have inserted between them a device whose input matches the output of the first circuit and whose output matches the input of the second circuit.

matched power gain—The power gain that is obtained when the output impedance of an amplifier is matched to the impedance of its load.

matching impedance—The impedance value that is needed across the terminals of a signal-voltage source for proper matching.

matching transformer—A transformer used for matching impedances.

matrix—In disc recording, a negative, metal copy of the master, used either as a stamper for pressing the final records, or as a pattern for a metal mother from which, in turn, a stamper can be made.

matrix circuit—A circuit for combining two or more signals in various proportions and phases. In quadraphonic sound systems it is used to combine the four channel signals into two channels on tape or disc and requires a matrix decoder circuit to recover the four signals in a reasonable approximation of their original form.

maximum output—The highest average output power that can be developed by a power source into a rated load, regardless of distortion.

maximum record level—The level of record-head current required to produce 3-percent third-harmonic distortion of the reproduced signal at the record-level set frequency. Distortion from electronic circuitry is not considered; only distortion from magnetic-tape saturation effects is measured.

maximum sound pressure—The maximum instantaneous value of sound pressure occurring during a complete cycle of a periodic wave. A common unit of measurement is the microbar.

maximum undistorted output—Also called maximum useful output. The maximum output power that can be obtained from an amplifier without producing excessive harmonics.

maximum useful output—*See* Maximum Undistorted Output.

maxwell—The unit of magnetic flux in the cgs system. It is equal to 1 gauss per square centimeter, or one magnetic line of force.

***m*-derived filter**—A wave filter whose formula allows the designer to control impedance and attenuation characteristics. The factor m in the formula is usually chosen as 0.60 for audio frequencies.

mean free path—The average distance that sound waves travel between successive reflections in an enclosure.

mechanical compliance—The ease with which a mechanical element can be displaced from its rest position. Compliance is the reciprocal of stiffness and compares to capacitance in an electrical circuit. It is usually stated in centimeters per dyne.

mechanical damping—A feature of a moving mechanism that tends to slow the movement. Damping is often accomplished by the use of some viscous fluid applied to the moving part.

mechanical impedance— The mechanical analog of electrical impedance. It is the ratio of effective force applied to an acoustic medium or mechanical device to the effective linear velocity produced by that force. In the cgs system the unit of impedance is the dyne-sec/cm. In the mks system the unit is the newton-sec/m.

mechanical ohm—The unit of mechanical impedance. The impedance for which a force of 1 dyne produces a linear velocity of 1 centimeter per second (dyne-sec/cm). When expressed in newton-sec/m it is called the mks mechanical ohm.

mechanical phonograph— A phonograph operating without the benefit of electrical pickup or amplification, but producing sound solely through the vibration of mechanical parts.

mechanical phonograph recorder—Equipment for converting electric or acoustic signals into mechanical motion and impressing same on a recording medium by cutting or embossing.

mechanical reactance—The imaginary component of mechanical impedance.

mechanical recorder— *See* Mechanical Phonograph Recorder.

mechanical recording head—*See* Cutter.

mechanical reproducer—*See* Phonograph Pickup.

mechanical resistance—The real part of the mechanical impedance. The cgs unit is the mechanical ohm. The mks unit is the mks ohm.

mechanical transducer—A device that transforms mechanical energy directly into acoustical energy.

meg—Abbreviation for megohm.

mega—Abbreviated M. Prefix denoting 10^6 (one million).

megabar—The absolute unit of pressure (equal to one million bars).

megahertz—One million hertz. Abbreviated MHz.

megohm—Abbreviated meg. One million ohms.

mel—A unit of musical pitch. A simple 1000-hertz tone 40 dB above the threshold of hearing produces a pitch of 1000 mels. The pitch of any sound that is judged by the listener to be *n* times that of a 1-mel tone is *n* mels.

metal film resistor—A resistor made by depositing an extremely thin layer of metal alloy on a nonconducting substrate.

metallized capacitor—A capacitor made by vacuum-depositing a thin film of metal directly on a dielectric, which can be paper or plastic. The capacitor tends to be self-healing; when a breakdown occurs, the metal film around it immediately burns away.

metal-oxide resistor—A type of film resistor in which the resistive material is tin oxide. One characteristic of this type is good stability.

mH—Abbreviation for millihenry.

mho—The unit of conductance. It is the reciprocal of an ohm. Therefore since the ohm equals 1 volt per ampere, the mho equals 1 ampere per volt. (Sometimes called siemens.)

MHz—Abbreviation for megahertz.

mica capacitor—A fixed capacitor that uses mica sheets as the dielectric material. The entire assembly is encased in a molded Bakelite or plastic case.

micro—In the metric system, a prefix meaning one millionth (1/1,000,000).

microbar—A unit of pressure commonly used in acoustics. One microbar is equal to one dyne per square centimeter. (In an earlier use, the term "bar" meant 1 dyne per square centimeter.)

microfarad—One millionth of a farad. Abbreviated mfd, mf, or μF.

microgroove—A term describing the narrow groove of most long-play and 45-rpm records. Normal width of the microgroove is 0.001 inch, about half as wide as the groove on the old 78-rpm record.

microgroove record— *See* Long-Play Record.

microhenry—One-millionth of a henry.

microhm—One-millionth of an ohm.

micromho—A unit of conductance equal to one-millionth of a mho.

micromicro—An obsolete prefix meaning one-millionth of a millionth, or 10^{-12}. Now called pico.

micromicrofarad—Obsolete term for 10^{-12} farad. Now called picofarad.

micromicrowatt—Obsolete term for 10^{-12} watt. Now called picowatt.

micron—1. A unit of length equal to 10^{-6} meter. 2. A unit of pressure measurement. One micron is equivalent to 0.001 mm (10^{-6} meter) of mercury at 32°F.

microphone—A transducer whose input is sound waves or audio vibrations and whose output is a corresponding electrical signal.

microphone boom—A movable crane for suspending a microphone before a sound source.

microphone amplifier—Also called a microphone preamplifier. An audio-frequency amplifier that boosts the output of a microphone to an appropriate level for the main audio-frequency amplifier.

microphone cable—A shielded cable for connecting a microphone to an amplifier.

microphone button—The button-shaped resistance element of a carbon microphone.

microphone preamplifier—*See* Microphone Amplifier.

microphone mixer—An audio mixer capable of combining the output from two or more microphones into a single input to an audio amplifier. The output level from each microphone is adjustable by individual controls on the mixer.

microphone stand—A stand for holding a microphone at an appropriate height before a performer. It may be designed for either floor or table use.

microphone transformer—An iron-core transformer used for matching the impedance of a microphone to an amplifier or transmission line.

microphonics—The generation of an electrical noise signal by mechanical motion of some part of a device. In its most common sense the term describes a form of acoustical feedback caused by vibration of some part of an electrical circuit by sound waves, thereby introducing an electrical signal in phase with the sound signal producing it.

microsecond—One-millionth of a second.

microvolt—One-millionth of a volt.

microwatt—One-millionth of a watt.

mike—Shortened word for microphone.

mil—One-thousandth of an inch. A unit in common use in the United States for measuring wire diameter.

milli—Abbreviated m. Prefix meaning one-thousandth (1/1000, or 10^{-3}).

milliampere—Abbreviated mA. One-thousandth (.001) of an ampere.

millihenry—Abbreviated mH. One-thousandth (.001) of a henry.

millimicro—Obsolete prefix for nano, equal to 10^{-9}.

millimaxwell—One-thousandth of a maxwell.

millisecond—One-thousandth of a second. Abbreviated ms.

millivolt—One-thousandth of a volt. Abbreviated mV.

milliwatt—One-thousandth of a watt. In audio, the reference level used for dB measurements.

minigroove—A recording having more lines per inch than the average 78-rpm phonograph record but not as many as the extended play, long playing, or microgroove records.

mismatch—A combination where the impedance of a load does not match the impedance of the source to which it is connected.

mitte-seite stereo system—German term for middle-side system. A system of stereo pickup in which

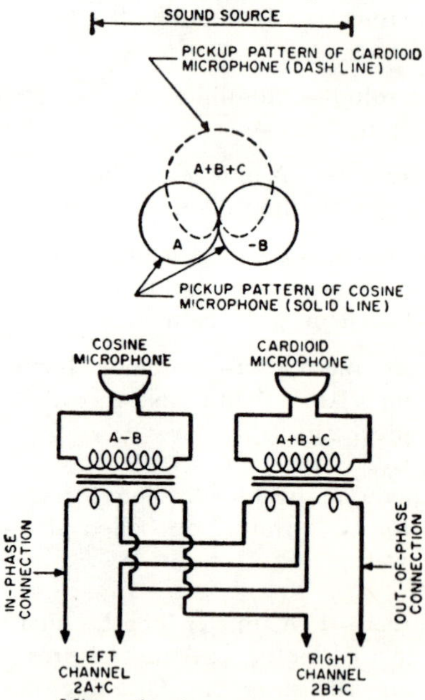

Mitte-seite stereo system.

two directional microphones are placed close together and at right angles to each other, then oriented so that one picks up sound from directly ahead, and the other from the two sides.

mixer—In a sound transmission, recording, or reproducing system, a device that will accept two or more inputs and combine them in a common output. The signals are combined linearly, hence one does not modulate the other.

mksa electromagnetic system of units—Also called the Giorgi system. A system in which the fundamental units are the meter, kilogram, second, and ampere.

mm—Abbreviation for millimeter.

mmf, mmfd—Abbreviation for micromicrofarad; now replaced by picofarad.

modulation—A process whereby a message or a form of intelligence is impressed on a wave by changing certain characteristcs of the wave such as its amplitude or frequency, or, if the wave has the nature of pulses, by varying the width, amplitude, or other aspect of the pulses.

modulation envelope—A curve drawn by connecting the peaks of a graph with a continuous line, showing how the waveform of a modulated carrier represents the waveform of the intelligence carried by the signal. The modulation envelope is the intelligence waveform.

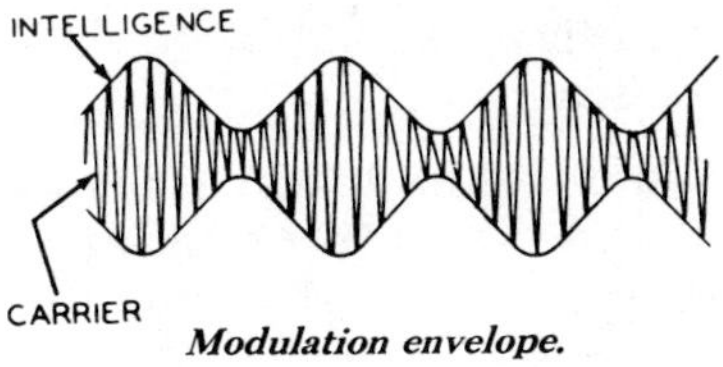

Modulation envelope.

modulation factor—In an amplitude-modulated wave, the modulation factor M is expressed by:

$$M = \frac{A_{max} - A_{min}}{A_{max} + A_{min}}$$

where,

A_{max} is the maximum value of the modulated wave.

A_{min} is the minimum value of the modulated wave.

modulation frequency—The frequency of the signal used to modulate an oscillator signal.

modulation noise—Noise caused by the signal, but not including any of the signal, itself.

modulation percentage— The modulation factor multiplied by 100.

modulator—A device used to impress modulation on a carrier frequency.

moisture absorption—A figure that indicates how much moisture (in percent) that an insulating material will absorb. If the insulation is to be used in a moist environment, the moisture absorption figure should be as low as possible.

moisture resistance—The ability of a material to resist moisture absorption, either from air or from immersion in water.

monaural—*See* Monophonic.

monaural recorder—Literally, the term would indicate a tape recorder designed for listening with one ear only; more correctly, the recorder in question should be called a monophonic recorder. The term is popularly used to refer to single-channel recorders as opposed to multichannel recorders.

monitor—To listen or otherwise observe a communication service to check its performance.

monitor head—On some tape recorders, an additional playback head that permits the operator to check recorded material as it is being recorded.

monitoring amplifier—An amplifier connected at some point in a listening or recording chain to evaluate and supervise results.

monogroove stereo—Also called single-groove stereo. A stereo recording process in which both channels are recorded on one groove.

monophonic—Also called by the older term "monaural." Literal meaning is "one sound" or "one ear." Refers to audio information on one channel (as opposed to stereophonic or binaural) with no attempt to represent spatial separation.

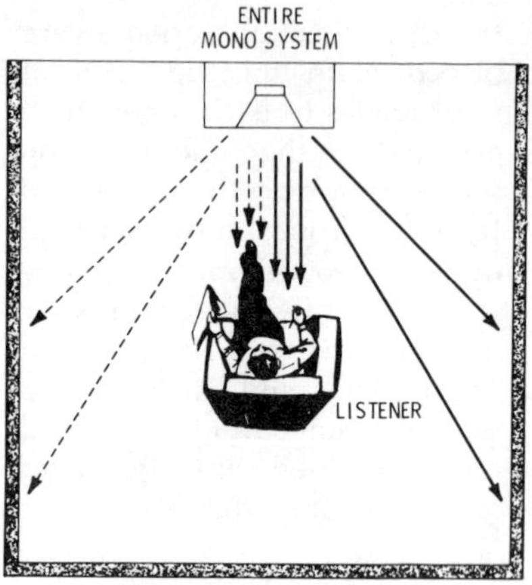

Monophonic system.

monophonic recorder—*See* Monaural Recorder.

monorange speaker—A single speaker used to provide the full spectrum of audio frequencies.

MOSFET—Abbreviation for metal-oxide-semiconductor field-effect transistor.

mother—In disc recording, a nickel-plated positive of the original record used for making stampers.

motional impedance—Also called loaded motional impedance. The impedance of a speaker when the voice coil is in motion (load impedance) minus the blocked impedance (impedance when the voice coil is blocked).

motor—Literally, a device for producing motion. Specifically, a machine that converts electric energy into mechanical energy.

motor board—Also called tape-transport mechanism. The structure of a tape recorder that serves as a mount for the motor (or motors) reel seats, heads, and controls. Usually includes all parts except amplifiers, speakers, and case.

motorboating—A form of low-frequency self-oscillation in an amplifier system, often originating in the power supply because of poor filtering or insufficient decoupling. So called because of resemblance to the familiar "putt-putt" of a motorboat.

mouth of a horn—The end having the larger cross section. The end opposite the throat.

moving-coil microphone—Also called a dynamic microphone. A microphone in which the diaphragm is attached to a coil positioned in a fixed magnetic field. When sound waves vibrate the diaphragm, the coil moves in the magnetic field, generating electrical signals in step with the sound waves.

moving-coil pickup—Also called dynamic pickup. A phonograph pickup in which the stylus vibrations cause a conductor or coil to move back and forth in a magnetic field, producing an electrical output.

moving-coil speaker—Also called a dynamic speaker. A speaker in which the moving diaphragm has an attached coil connected to the signal source and placed in a constant magnetic field. Current through the coil produces a magnetic field that interacts with the permanent field of the speaker to produce movement of the diaphragm.

moving-conductor microphone—A microphone that generates an electrical signal by the motion of a conductor in a magnetic field.

ms—Abbreviation for millisecond.

msec—Abbreviation for millisecond.

mu—English spelling for the Greek letter μ.

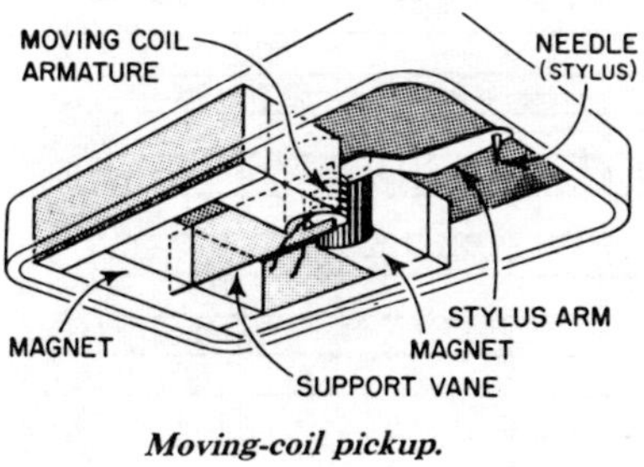

Moving-coil pickup.

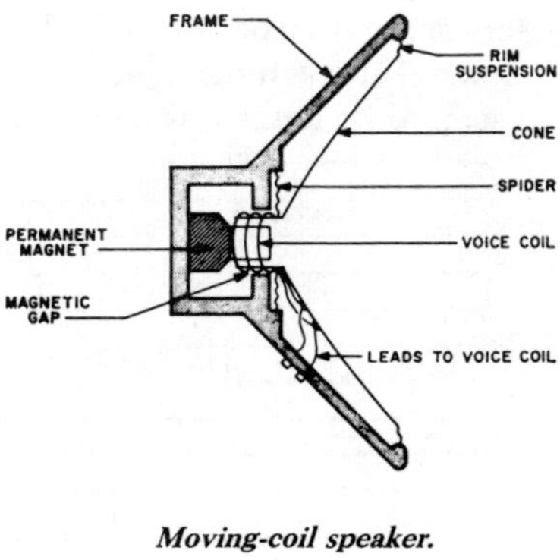

Moving-coil speaker.

μ—Greek letter mu. 1. Symbol for amplification factor. 2. Symbol for permeability. 3. Abbreviation for micron. 4. Abbreviation for the prefix micro.

μA—Abbreviation for microampere.

μH—Abbreviation for microhenry.

μmho—Abbreviation for micromho.

μΩ—Abbreviation for micro-ohm.

μs—Abbreviation for microsecond.

μV—Abbreviation for microvolt.

μW—Abbreviation for microwatt.

multicasting— Broadcasting a stereo program by using two fm stations. Reception of this type of broadcast requires two fm receivers.

multicellular horn—An assembly of a number of exponential horns so arranged that vertical and horizontal dispersion is controlled.

multipath reception—Arrival of radio signals by several paths of different lengths, because of reflection by intervening objects. Such reception causes problems because of phase differences in the received signals.

multiple sound track—Two or more sound tracks recorded side by side on the same medium, to be played back at the same time (for example, a stereophonic recording).

multiplex—To send several messages or signals simultaneously on the same circuit or channel.

multiplex adapter—An fm tuner or receiver designed to receive two-channel or stereophonic signals from a station transmitting multiplex broadcasts. The right and left signals are separated by the demodulation process.

multiplex stereo—A system of broadcasting a stereo program by sending both channels on a single carrier. This is accomplished by modulating an ultrasonic subcarrier—either with the signals of one of the stereo channels or by a difference signal obtained by combining the two channels out of phase.

multitrack magnetic system—A recording system in which two or more tracks are placed on the magnetic recording medium.

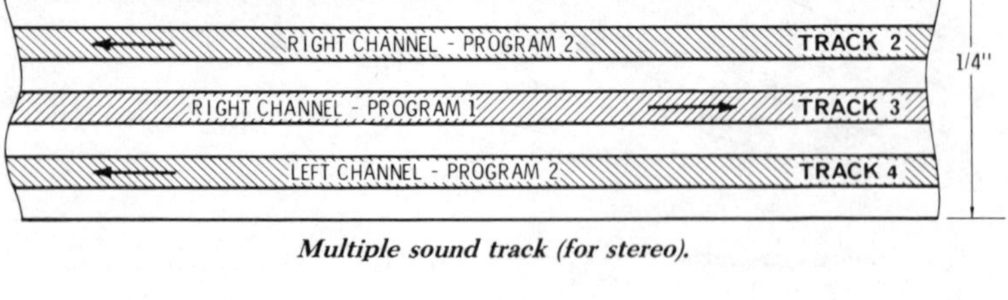

Multiple sound track (for stereo).

multitrack recording system—A recording system with two or more recording paths designed for playback at the same time.

musical echo—*See* Flutter Echo, 2.

musical quality—*See* Timbre.

music power—The power delivered by an amplifier when reproducing short-term peaks of program material. The music power rating exceeds the rms power rating, more or less, and was standardized by the Institute of High Fidelity as being more representative of amplifier capabilities when reproducing program material.

muting—The muffling or deadening of a sound.

mutual impedance—The mutual impedance between any two pairs of network terminals (all other terminals being open) can be found by dividing the open-circuit potential at either one of the two pairs by the current applied to the other pair.

mutual inductance—A property that exists between two current-carrying conductors in a magnetic field common to both. It determines, for a given rate of change of current in one circuit, the electromotive force induced in the other.

mutual induction—The property by which a changing current in one circuit induces a changing voltage in a neighboring circuit, although no direct connection exists between the two circuits.

mV—Abbreviation for millivolt.

Mylar—Trade name of E.I. duPont de Nemours and Co., Inc., for a transparent plastic film of high durability and outstanding strength. It is widely used as a base for magnetic tape and as a dielectric in capacitors.

N

n—Abbreviation for the prefix nano-(10^{-9}).

nA—Abbreviation for nanoampere.

NAB—Abbreviation for National Association of Broadcasters.

NAB curve—A standard playback equalization curve set by the National Association of Broadcasters.

nano—Abbreviated n. Prefix meaning one-billionth (10^{-9}).

nanoampere—One-thousandth of a microampere (10^{-9} ampere). Abbreviated nA.

nanofarad—One-billionth of a farad, equal to 10^{-9} farad, 0.001 microfarad, and 1000 picofarads. Abbreviated nF.

nanohenry—One-thousandth of a microhenry, equal to 10^{-9} henry and 1000 picohenrys. Abbreviated nH.

nanosecond—One-billionth of a second (10^{-9} second). Abbreviated ns.

nanovolt—One-thousandth of a microvolt (10^{-9} volt. Abbreviated nV.

narrow band—1. A band whose width is greater than 1% of the center frequency and less than 1/3 octave. 2. Term descriptive of a communication channel not

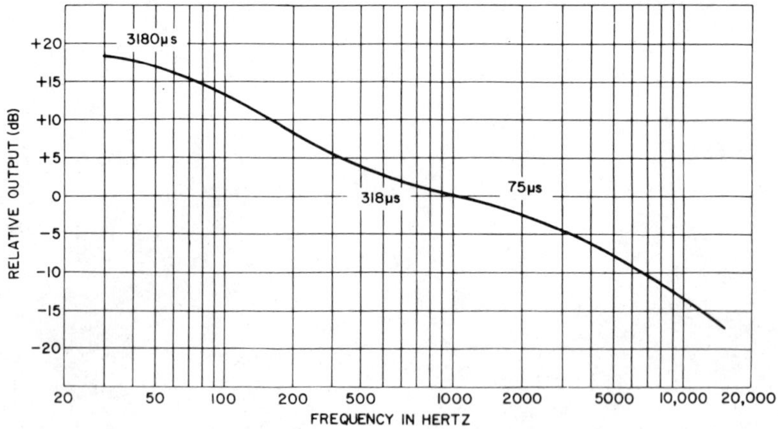

NAB curve (RIAA standard reproducing characteristic).

wide enough for good voice transmission.

narrow-band amplifier—An amplifier purposely designed to operate over a narrow band of frequencies.

NARTB—Abbreviation for National Association of Radio and Television Broadcasters.

National Association of Broadcasters—Abbreviated NAB. An association of radio and television broadcasters.

National Association of Radio and Television Broadcasters—Abbreviated NARTB. A name used for a number of years by an association of broadcasters.

natural frequency—1. The frequency at which an elastic body will oscillate if subjected to a momentary displacement. 2. The lowest resonant frequency of a circuit or component obtained without adding inductance or capacitance.

natural magnet—Magnetic ore that when found in its natural state exhibits magnetic properties. Sometimes called lodestone (leading stone).

natural period—The period of vibration of an oscillating body or system when momentarily disturbed and then allowed to oscillate without further influence. In systems whose period varies with amplitude, the natural period is the period as the amplitude approaches zero.

natural resonance—*See* Periodic Resonance.

nbfm—Abbreviation for narrow-band frequency modulation.

near field—The acoustic radiation field near the source of sound (speaker, etc.).

needle—*See* Stylus.

needle chatter—*See* Needle Talk.

needle drag—*See* Stylus Drag.

needle pressure—*See* Stylus Force.

needle scratch—*See* Surface Noise.

needle talk—Also called needle chatter. Sounds produced by direct radiation from the surface of the vibrating needle (stylus) of a phonograph pickup.

negative electrode—The element of a primary cell that supplies electrons to the negative terminal of the cell and to any external circuit connected to it.

negative feedback—Also called inverse feedback, degeneration, and stabilized feedback. The application of part of the output of an amplifier to some preceding point in the amplifier circuit where the signal is 180° out of phase with the output. The amplification is thereby decreased and distortion is reduced.

negative-feedback amplifier—An amplifier that uses negative feedback to improve the stability or frequency response, or both.

negative impedance—If the reactive elements (inductance and capacitance) are absent it is then called negative resistance. A characteristic of certain electrical devices or circuits—when voltage increases, current decreases, and vice versa.

negative resistance—A resistance whose voltage and current characteristics are the inverse of Ohm's law—when the voltage increases the current decreases and vice versa.

negative-resistance region—That operating region in which an increase in applied voltage results in a decrease in current.

negative temperature coefficient—A change in the value of a quantity, such as capacitance or resistance, that is opposite in direction to the change in temperature that produces it; i.e., an increase in temperature causes a decrease in the value of the quantity.

NEMA—Abbreviation for National Electrical Manufacturers Association.

NEMA standards—Specifications adopted as standard by the National Electrical Manufacturers Association.

neper—also called napier—A logarithmic unit used for expressing ratios between voltages, currents, or powers. Similar to the decibel, but based on natural logarithms rather than common logarithms. One neper equals 8.686 decibels. The ratio between two voltages is expressed by the formula:

$$N = \log_{\epsilon} \sqrt{\frac{E_1}{E_2}}$$

where,

N is the number of nepers denoting their ratio,

E_1 and E_2 are the two voltages.

net reactance—The difference between the capacitive and inductive reactances in an ac circuit.

network—A combination of electrical elements connected in a circuit.

network filter—A combination of circuit elements for separating waves on the basis of their frequencies.

newton—A unit of force in the mksa system of physical units. The unit of force that will impart an acceleration of 1 meter per second per second to a mass of 1 kilogram.

nf—Abbreviation for noise figure.

nF—Abbreviation for nanofarad.

nif—Abbreviation for noise-improvement factor.

node—A stationary point in a vibrating string or diaphragm.

noise—1. Unwanted, unpleasant sound lacking in musical quality and having no predominant frequency. 2. An unwanted electrical disturbance that interferes with the transmitting, indicating, or recording of a desired signal. 3. Random electrical signals generated by mechanical or thermal disturbances in some portion of an amplifier or transducer.

noise analysis—Determination of the frequency components that make up the noise being studied.

noise analyzer—An instrument used for determining the frequency and amplitude characteristics of noise.

noise behind the signal—*See* Modulation Noise.

noise-cancelling microphone—*See* Close-Talking Microphone.

noise-criteria value—A term used to indicate the level of background noise. The value is derived from the highest levels of sound pressure of several frequency bands. Also called NC value.

noise level—1. The level in dB of unwanted sounds at a given location. 2. The strength of unwanted signals in a circuit. (Noise level is usually measured in dB with respect to a certain reference level.)

noise suppression—1. A feature of some radio receivers that reduces the noise output when no carrier is being received. 2. A means of reducing surface noise during reproduction of a phonograph record.

noise suppressor—1. A circuit used in some record players for reducing hiss and noise when old or damaged records are played. 2. In a receiver circuit, the portion that reduces noise automatically when no carrier is being received.

nondirectional—*See* Omnidirectional.

nondirectional microphone—*See* Omnidirectional Microphone.

nonlinear—Literally, not straight. A general term describing a condition where a curved-line (nonlinear) graph is obtained when cause and effect, input and output, etc., are plotted.

nonlinear distortion—Distortion that occurs when the output does not rise and fall in direct proportion to the input. The input and output values need not be of the same physical quantity—e.g., in a linear detector, these values are the signal voltage at the output and the modulation envelope at the input.

nonmagnetic—Material that has no appreciable magnetic properties; that is, it cannot be magnetized or attracted by a magnet.

nonmagnetic steel—A steel alloy that contains about 12% manganese, and sometimes a small quantity of nickel. It is practically nonmagnetic at ordinary temperatures.

nonsinusoidal wave—Any wave that is not a sine wave. Therefore, it does not consist of a single frequency, but of several.

nonsynchronous—Not having the same period or phase as other quantities in a device or circuit.

normal impedance—*See* Free Impedance.

normal induction curve—The curve obtained by plotting B (induction) against H (magnetizing force), starting from a totally demagnetized state.

normal record level—In direct recording, the record-head current that will produce 1% third-harmonic distortion of the reproduced signal at the record-level-set frequency. This is distortion produced by magnetic-tape saturation, not by electronic circuitry.

north pole—The end of a magnet from which magnetic lines of force are considered to leave; the north-seeking pole of a compass.

note—1. A sensation of tone. 2. A single musical tone having position on a musical scale (A, C, G, etc.) 3. A printed character used to indicate a specific tone and its duration.

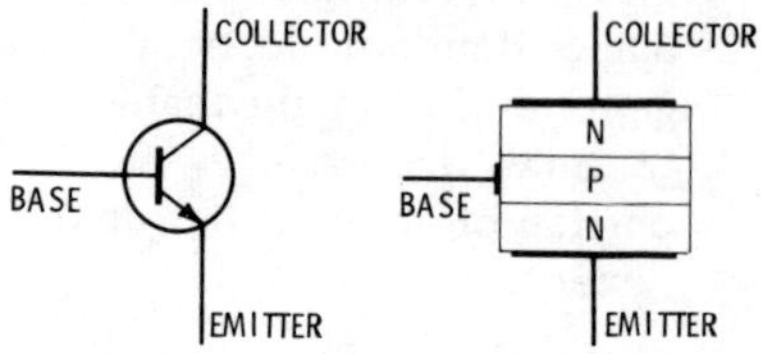

Symbol. *Junctions.*

Npn Transistor.

npn transistor—A transistor with a p-type base and an n-type collector and emitter.

ns—Abbreviation for nanosecond (10^{-9} sec).

nth harmonic—A harmonic having a frequency *n* times the frequency of the fundamental component.

nV—Abbreviation for nanovolt (10^{-9} volt).

nW—Abbreviation for nanowatt (10^{-9} watt).

O

octave—1. A musical interval between two tones that have a basic frequency ratio of 2 to 1. 2. In a continuous musical scale, the group of notes lying between any given note and the next note above or below in pitch, of the same name (e.g., A through G).

octave band—A continuous band of frequencies with a 2 to 1 ratio of upper limit to lower limit.

odd harmonic—Any frequency that is an odd multiple (3, 5, 7, etc.) of the fundamental frequency.

oersted—A unit of magnetizing force in the cgs electromagnetic system. It is equal to $1000/4\pi$ ampere-turns per meter. At any point in a vacuum, the magnetic intensity in oersteds is equal to the force in dynes exerted on a unit magnetic pole at that point.

offset angle—In lateral-disc reproduction, the smaller of the two angles between a line through the arm pivot point and the stylus pivot point, and a line through the stylus tip and the stylus pivot point, all points being projected into the plane of the disc. The stylus is assumed to be in its rest position.

ohm—The unit of resistance, symbolized by the Greek letter omega (Ω). It is defined as the resistance at 0°C of a uniform column of mercury 106.300 cm long and weighing 14.4521 grams. One ohm is the resistance through which a potential difference of one volt will maintain a current of one ampere.

ohmic contact—A contact having a resistance that obeys Ohm's law—that is, a contact across which the potential difference is proportional to the current passing through it.

ohmic resistance—Resistance to direct current according to Ohm's law.

Ohm's law—In a given conductor at a given temperature, the current is directly proportional to the difference in potential between the ends of the conductor. Expressed mathematically, this takes the form:

$$I \propto E$$

The proportionality constant is written $1/R$, where R is called the resistance. The formula then becomes $I = E/R$, or by transposition, $E = IR$ and $R = E/I$. These last three forms are the commonly used expressions for Ohms law.

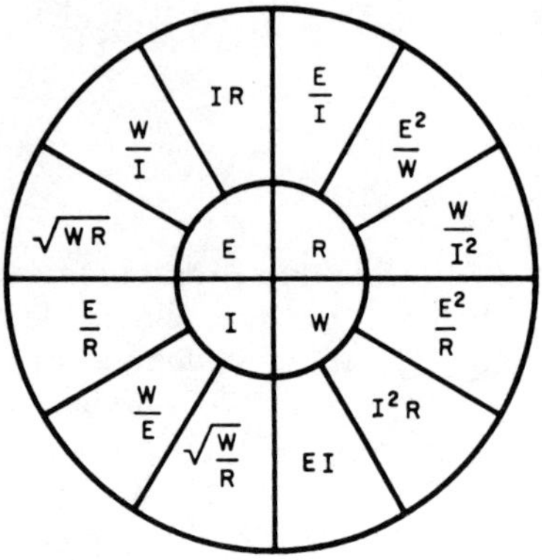

Ohm's law.

omnidirectional—Also called nondirectional. All-directional; not favoring any one direction.

omnidirectional microphone—Also called nondirectional microphone. A microphone that responds to sound waves from all directions.

open-circuit jack—A jack that leaves its circuit open when the plug is removed. The circuit is closed when the plug is inserted.

open-circuit voltage—The voltage at the terminals of a battery or other voltage source when no load current is being drawn.

open-loop gain—The gain of an amplifier that normally has feedback, if the feedback is disabled; i.e., if the feedback loop is opened.

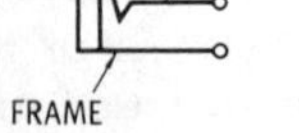

Open-circuit jack.

open magnetic circuit—A magnet that does not have a complete path of ferromagnetic material, either through the magnet itself, or through magnet and external materials.

operating cycle—The complete sequence of operations required in the normal functioning of an item of equipment.

operating point—Also called quiescent point. A point on a family of characteristic curves (either vacuum tube or transistor) that represents instantaneous electrode currents and voltages for the chosen operating conditions.

opposition—The phase relationship between two periodic functions of the same period when the phase difference is 180° (half a period).

optical pattern—Also called Christmas-tree pattern. In disc recording, the pattern observed when the surface of the record is illuminated by a light beam of essentially parallel rays. The shape of the pattern is used to check the recording characteristic.

optical sound—A system of recording and reproducing sound by using modulated light areas near one edge of a motion-picture film.

optimum damping—A value of damping about 65% of critical damping. It permits fast re-

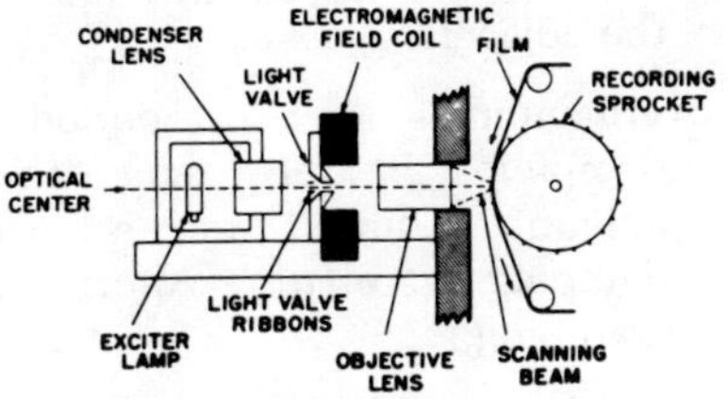

Optical sound recorder.

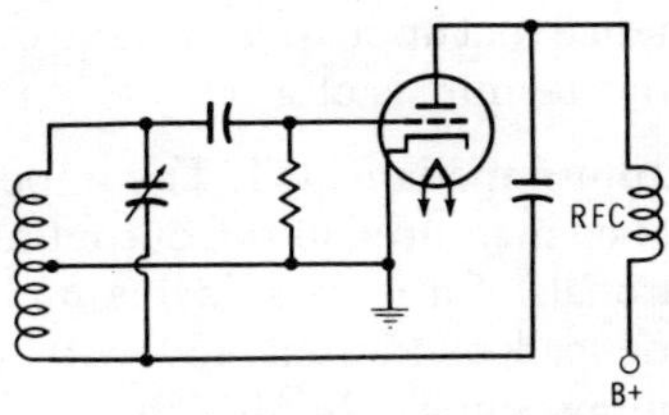

Oscillator (shunt-fed Hartley).

sponse but also allows a little overshoot.

optimum load—The value of load impedance that will transfer maximum power from the source to the load.

orange peel—A descriptive term for the surface of a recording blank that resembles an orange peel in texture. Such a surface has a high background noise.

original lacquer—An original lacquer disc recording made for the purpose of producing a master.

original master—In disc recording, the metal master (negative) produced by electroforming from the face of a wax or lacquer recording.

oscillation—1. In electricity, a flow changing periodically from maximum current to minimum. 2. Periodic vibration in a mechanical system. 3. A single vibration or alternation.

oscillator—An electrical or mechanical device for producing oscillations. In electronics, usually characterized by an amplifier with positive feedback.

oscillograph—An instrument used primarily for recording the instantaneous values of one or more electrical quantities that vary as a function of time, or of another electrical or mechanical quantity.

oscilloscope—An instrument used primarily for displaying the instantaneous values of one or more rapidly varying electrical quantities as a function of time, or of another electrical or mechanical quantity. It does not inherently produce a permanent record but is often combined with a camera for that purpose.

out of phase—Said of two or more waveforms that pass through corresponding values, but not at the same instant.

outphasing—In electronic organs, a term applied to a method sometimes used for producing certain voices. In this method, harmonics or subharmonics of the tone-generator signal are added or subtracted to the circuit at a point between the keying-system output and the formant filters.

output capacitance—The capacitance that appears across the output terminals of a device.

output impedance—1. The impedance measured at the output terminals of a power device under no-load conditions and with no input signal. 2. Also called dynamic output impedance. The impedance presented to the load by a power supply. It can be calculated as the ratio of change in output voltage to the change in load current causing the change.

output power—The power that a system or component delivers to its load.

output stage—The final stage in any electronic equipment; the stage from which output is taken. In a radio receiver, it feeds the speaker directly or through an output transformer. In an audio-frequency amplifier, it feeds one or more speakers, the cutting head of a sound recorder, a transmission line, or any other load.

output transformer—A transformer used to couple and match the output stage of an amplifier to a load.

output voltage—The maximum output voltage that an amplifier will develop in the linear operating region (within acceptable limits of distortion).

overcutting—In disc recording, the application of too strong a signal to the cutting head, resulting in one groove being cut through to the adjacent groove.

overdamping— Any periodic damping greater than the amount required for critical damping. *(See also* Aperiodic Damping.)

overdriven amplifier—An amplifier stage driven to distortion, either intentionally or unintentionally, by applying too large a signal for its normal operating characteristics.

overload—A load greater than that which an amplifier, other component, or an entire system is designed to carry. It is usually accompanied by waveform distortion or overheating and may result in destruction of components.

overload capacity—The maximum level of current, voltage, or power that a device will stand without being ruined. It is usually higher than the rated load capacity.

overload protection—A device that automatically disconnects the circuit or limits circuit loading whenever the current or voltage becomes excessive.

overmodulation—Overloading or exceeding the maximum amplitude for a given recording system. In radio a-m broadcasting, modulation with a signal that exceeds the amplitude of the carrier signal. The carrier is therefore reduced to zero at these values of signal.

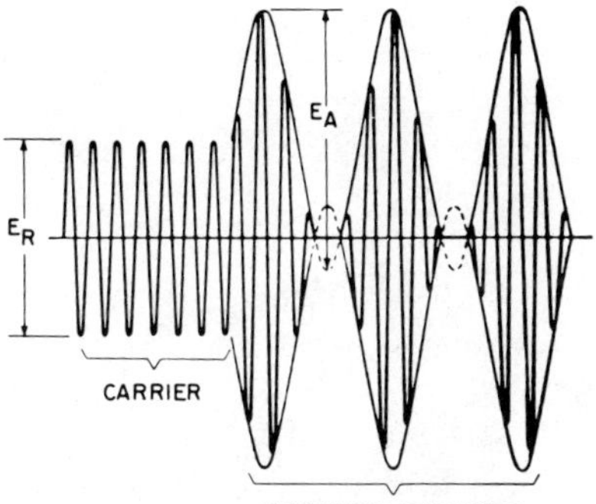

Overmodulation.

overshoot—1. The initial response that exceeds the steady-state response when a step signal or steep wavefront is suddenly applied to the input. 2. Amplitude of the first maximum excursion of a pulse beyond the 100% amplitude level, expressed as a percentage of this 100% amplitude level. 3. In a system with power regulation, a transient rise beyond regulated output limits, occurring when the ac power is turned on or off, and for step changes of line or load.

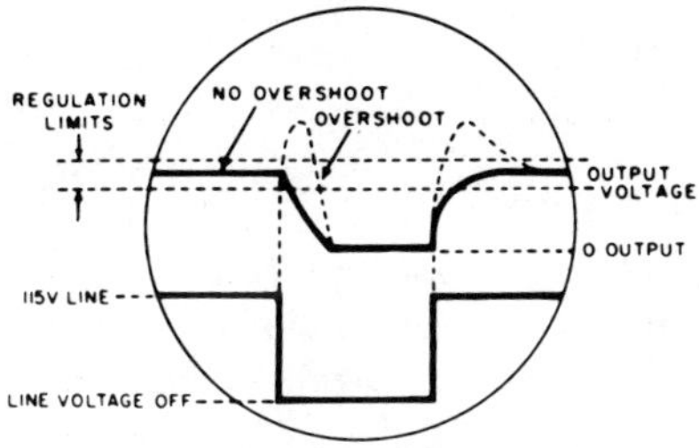

Overshoot, 3.

overthrow distortion—Also called overshoot distortion. The distortion that occurs in a signal wave when the maximum amplitude of the signal wavefront exceeds the steady-state amplitude.

overtone—A component of a complex tone with a pitch higher than the fundamental. Sometimes confused with "harmonic."

oxide—In magnetic recording, one of various oxides used as the magnetic medium with which the recording tape is coated. Ferric oxides have been used extensively, and, more recently, chromic oxide.

P

p—Abbreviation for power (combining form, as in pf for power factor) or the prefix pico (10^{-12}).

P—Symbol for permeance.

pA—Abbreviation for picoampere.

packing—Crowding of the carbon particles in a carbon microphone. This affects the operation by changing the resistance from its normal value and reducing the sensitivity of the microphone.

pad—1. A component capable of reducing the amplitude of a wave without introducing appreciable distortion. 2. A device inserted into a circuit to introduce transmission loss or to match impedances.

pam—Abbreviation for pulse-amplitude modulation.

panoramic sonic analyzer—Also called panoramic spectrum analyzer. A heterodyne-type instrument that separates the frequency components of a complex waveform and displays them on an oscillographic screen, indicating both frequency and magnitude. Useful for checking distortion, equalizer and filter characteristics, vibration, noise, etc.

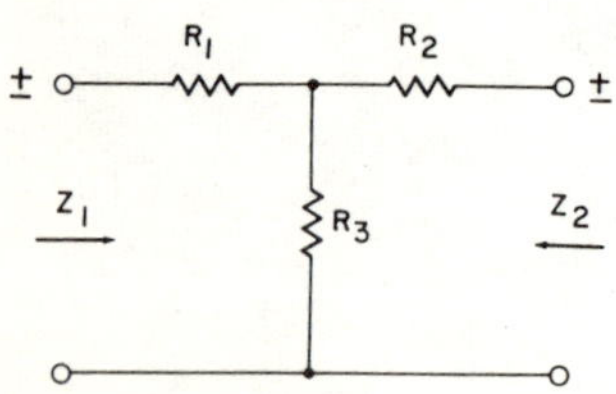

Pad, 2.

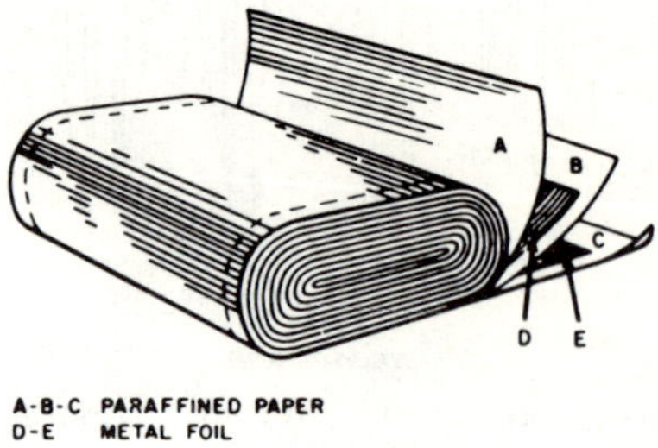

Paper capacitor.

paper capacitor—A fixed capacitor consisting of two strips of metal foil separated by oiled or waxed paper or other insulating material. A terminal is attached to each metal strip.

parabolic microphone—A microphone mounted at the focus of a parabolic sound reflector to obtain highly directional characteristics.

parallel circuit—A circuit in which all positive terminals connect to a common point and all negative terminals connect to another common point. The voltage is

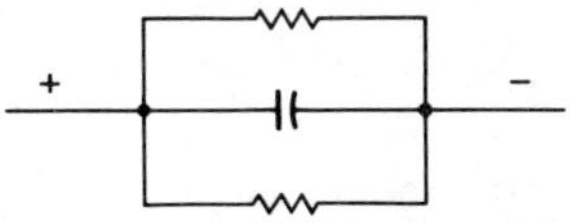

Parallel circuit.

the same across each element in the circuit.

parallel resonance—In a circuit comprising inductance and capacitance connected in parallel, a condition when the impressed frequency is such that the impedance of the parallel combination is a pure resistance, with zero phase angle.

parallel-resonant circuit—An inductor and a capacitor connected in parallel to furnish a high impedance at the frequency to which the circuit is resonant.

paramagnetic—Material is said to be paramagnetic if when placed in a magnetic field it becomes magnetized in the same direction as the applied field.

paraphase amplifier—An amplifier that converts a single input signal into two out-of-phase signals for driving a push-pull stage.

parasitic oscillation—1. An undesired, self-sustaining oscillation at a frequency other than

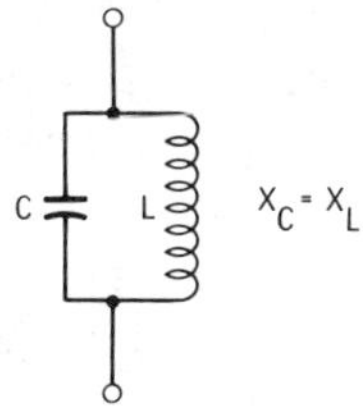

Parallel-resonant circuit.

the operating frequency. 2. Any undesired oscillation in an oscillator or amplifier stage.

parasitic suppressor—A parallel resistance, or a parallel combination of inductance and resistance, inserted into a grid or plate circuit to suppress parasitic oscillations.

partial—1. A physical component of a complex tone. 2. A component of a sound sensation that can be distinguished as a simple tone that cannot be further analyzed by the ear and that contributes to the character of the complex sound.

partial node—The place in a standing-wave system at which some characteristic of the wave field has a minimum amplitude other than zero.

particle velocity—The velocity of a very small particle or molecule of the sound transmission medium (usually air) as it is being displaced back and forth at the frequency of the sound wave. Usually measured in centimeters per second.

passband—The band of frequencies that will pass through a filter with very little attenuation.

passive component—A component of a system that contributes no power to the system but usually creates some loss.

passive device—A device that contributes no power gain to a circuit or system. It has no control action and does not require any input other than a signal to perform its function. Examples of passive devices are conductors, resistors, and capacitors.

passive network—A network with no source of energy.

pa system—Abbreviation for public-address system.

patch cord—Sometimes called an attachment cord. A short cord with a plug or pair of clips on one end for conveniently connecting two items of sound equipment such as a phonograph and tape recorder, an amplifier and speaker, etc.

pause control—A feature of some tape recorders, usually a button or slide switch with spring return, used to make a quick, temporary stop of tape movement without changing the position of the play or record controls.

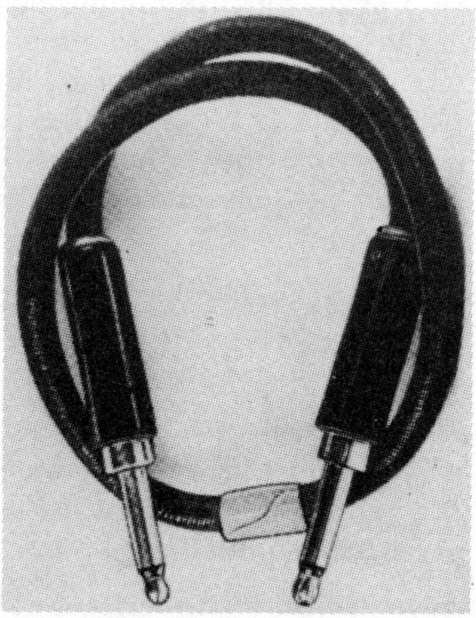

Patch cord.

pcm—Abbreviation for pulse code modulation.

pcm/fm—Frequency modulation of a carrier by pulse-code-modulated information.

pdm—Abbreviation for pulse-duration modulation.

pdm/fm—Frequency modulation of a carrier by pulses that are modulated in duration by information.

peak—The maximum instantaneous value of a quantity.

peak distortion—The greatest total distortion of signals occurring during a given period.

peak inverse voltage—The peak ac voltage that a rectifying cell or pn junction will withstand in the reverse direction.

peak level—The maximum instantaneous level that occurs during a specific time interval (i.e., in acoustics, the peak sound-pressure level).

peak pulse amplitude—The peak value of a pulse, excluding unwanted portions such as spikes.

peak pulse power—The maximum power of a pulse, excluding spikes.

peak response—The maximum response of a system to an input.

peak sound pressure—The maximum absolute value of instantaneous sound pressure for any specified time interval. The most common unit is the microbar.

peak-to-peak—The algebraic difference between the positive and negative maximum values of a waveform.

peak value—The maximum instantaneous value of a varying current, voltage, or power. For a sine wave, it is equal to 1.414 times the effective value of the sine wave.

peak voltage—The maximum value present in a varying or alternating voltage. This value may be either positive or negative.

pedal clavier—The pedal keyboard of an organ. It supplies the bass accompaniment for the other manuals.

perceived noise level—An empirical measure that includes allowance for the subjective reaction of people to noise in the various frequency ranges. It is expressed in decibels (PndB).

percent of harmonic distortion—A measure of the harmonic distortion in a system or component. It is equal to 100 times the ratio of the total rms voltage of all harmonics to the total rms voltage of harmonics and fundamental.

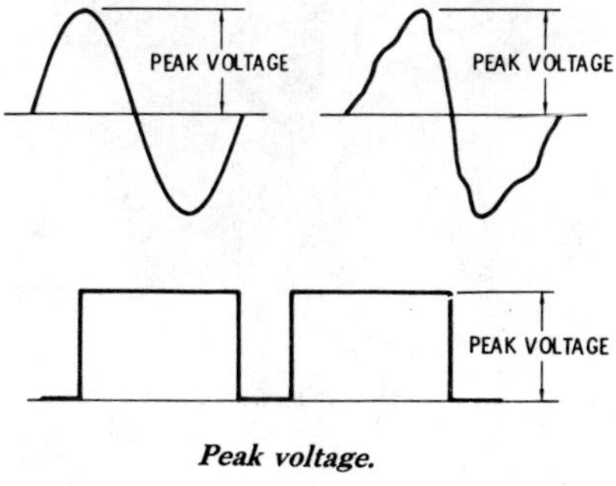

Peak voltage.

percent of modulation—1. In amplitude modulation, the ratio of half the difference between the maximum and minimum amplitudes of a wave to the average amplitude, expressed in percentage. 2. In frequency modulation, the ratio of the actual frequency swing to the frequency swing that is defined as 100% modulation, expressed in percentage. For fm broadcast stations, a frequency swing of ±75 kHz is defined as 100% modulation. For television, it is ±25 kHz.

percent of ripple voltage—Ratio of the effective (root-mean-square) value of the ripple voltage to the average value of the total voltage, expressed in percent.

percussion—1. Creation of sound by striking an object a sharp blow. 2. A class of musical instruments played by striking the vibrating member. The waveshape of the sound produced is characterized by a steep wavefront followed by a more or less rapid decay, depending on the damping qualities of the instrument. A few examples of percussion instruments are the piano, drums, xylophone, chimes, bells, tambourine, and cymbals.

period—The time required for one complete cycle of a regular, repeating series of events.

periodic—A phenomenon that repeats regularly in time and form.

periodic damping—Also called underdamping. Damping in which the pointer of an instrument oscillates about the final position before coming to rest. If the damping is increased enough, eventually the damping becomes aperiodic or overdamped. The point at which damping changes from periodic to aperiodic is called critical damping.

periodic pulse train—A pulse train made up of identical groups of pulses repeated at regular intervals.

periodic resonance—Also called natural resonance. Resonance in which the applied agency maintaining oscillation has the same frequency as the natural period of oscillation of a system.

periodic vibration—A vibration having a regularly recurring waveform; e.g., sinusoidal vibration.

periodic wave—A wave in which the displacement has a periodic variation with time, distance, or both.

permalloy—A high-permeability magnetic alloy composed mainly of iron and nickel.

permanent-field synchronous motor—A type of synchronous motor that has permanent-magnet field poles. It is constructed to have the starting characteristics of an induction motor and to operate at synchronous speed.

permanent magnet—A piece of hardened steel or other magnetic material of high retentivity that has been strongly magnetized and that retains the magnetism indefinitely.

permanent-magnet material—Ferromagnetic material that, once magnetized, resists external demagnetizing forces (i.e., requires a high coercive force to remove the magnetism).

permanent-magnet speaker—A moving-conductor speaker in which the steady magnetic field is produced by a permanent magnet.

permeability—Symbolized by the Greek letter *mu* (μ). The measure of the ease with which magnetic lines of force can be passed through a material, as compared

Permanent-magnet speaker.

to air. (Air is assumed to have a permeability of 1.) It is equal to the magnetic induction (B) in gausses, divided by the magnetizing force (H) in oersteds.

permeability tuning—A method of tuning by moving a magnetic core into or out of a coil to vary its inductance.

permeameter—An apparatus for determining the magnetizing force and flux density in a test specimen. The values obtained can be used to plot hysteresis loops and compute magnetic permeability.

permeance—The reciprocal of reluctance. In a magnetic circuit, it is the ratio of flux to magnetic potential difference. In the cgs system, it is equal to magnetic flux in maxwells divided by magnetomotive force in gilberts.

perpendicular magnetization—In magnetic recording, magnetization that is perpendicular to the line of travel and parallel to the smallest cross-sectional dimension of the medium. (*See also* Transverse Magnetization.)

pF—Abbreviation for picofarad.

pg—Abbreviation for power gain.

phantom channel—In a stereo system, an electrical combination of the left and right channels fed to a third, centrally located speaker.

phase—1. A time relation between two or more quantities that can be represented by the same waveform with a one-to-one correspondence between successive points, such points not necessarily occurring at the same instant of time. If the points of any two such waveforms do occur at the same time, the waveforms and the quantities they represent are said to be in phase; if they do not, the waveforms are said to be out of phase. For example, the voltage across a pure resistance is in phase with the current through it; the voltage across an inductor is out of phase with the current through it.

phase angle—1. Of a periodic function, the angle obtained by multiplying the phase (the elapsed fraction of the period) by 2π, if the angle is to be expressed in radians, or by 360, if the angle is to be expressed in degrees. 2. The angle between the vectors representing two periodic functions that have the same frequency. 3. The phase difference, in degrees, between corresponding stages of progress of two cyclic operations.

phase characteristic—A graph showing phase shift versus frequency for a sinusoidal input and output.

phase corrector—A network designed to correct for phase distortion.

phase difference—The time in electrical degrees by which one wave leads or lags another.

phase inversion—The action of a circuit that produces a negative-going output from a positive-going input and vice versa. A sine-wave input to a phase inverter produces a sine-wave output 180° out of phase with the input.

phase inverter—1. A stage that functions chiefly to change the phase of a signal by 180°, usually for feeding one side of a following push-pull amplifier. 2. *See* Vented Baffle.

phase meter—Also called phase-angle meter. An instrument for measuring phase difference between two quantities of the same frequency.

phase-modulated wave—The output of a carrier whose phase has been caused to deviate from its original (no-signal) angle by an amount proportional to the modulating signal amplitude.

phase modulation—Abbreviated pm. Modulation in which the angle of a sine-wave carrier deviates from the original (no-signal) angle by an amount proportional to the instantaneous value of the modulating wave. Phase and frequency modulation in combination are commonly referred to as "frequency modulation."

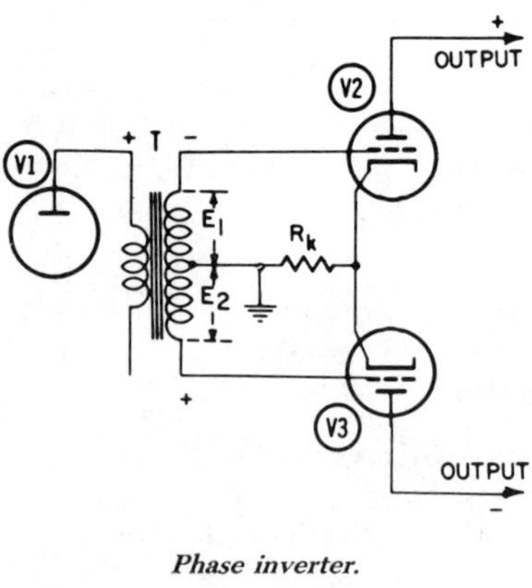

Phase inverter.

phase modulator—A circuit that modulates the phase of a carrier signal.

phase-response characteristic—The phase displacement versus frequency properties of a network or system.

phase reversal—A change in phase of 180°, or one-half cycle.

phase-reversal switch—A switch on a stereo amplifier or in a speaker system used to shift the phase of one channel by 180°.

phase shift—1. The time difference between corresponding points on input and output signal waveshapes of an ac signal, expressed as degrees of lead or lag. 2. A change in the phase of a periodic quantity.

phase-shift circuit—A network that shifts the phase of one ac quantity with respect to another ac quantity of the same frequency.

phase shifter—A device in which the output voltage (or current) may be adjusted to have some desired phase relationship with the input voltage (or current).

phase-shift microphone—A microphone having directional properties that are provided by phase-shift networks.

phase-shift oscillator—An oscillator in which a network having a phase shift of an odd multiple of 180° (per stage) at the oscillation frequency is connected between the output and the input of an amplifier. The 180-degree shift of the network, combined with an equal shift in the amplifier stage, produces a total shift of 360°, so that the feedback signal is in phase with the input signal, producing oscillation. When the phase shift is obtained by resistance-capacitance elements, the circuit is called an RC phase-shift oscillator.

phase splitter—A device that produces, from a single input wave, two or more output waves that differ in phase from one another.

phasing—1. Causing two systems or circuits to operate either in phase or at some desired difference from an in-phase condition. 2. In stereo application, or in multiple speaker systems, the connection of amplifier output to the speakers so that all speaker cones move in the same direction (either in or out) for a given polarity of applied voltage.

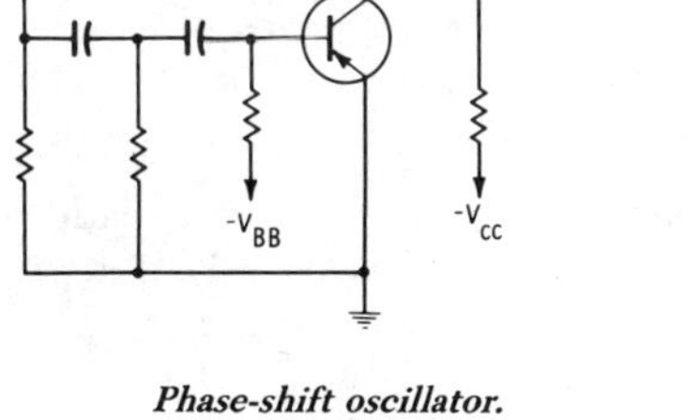

Phase-shift oscillator.

Phone jack.

phon—The unit of measurement for apparent loudness level of a sound. Numerically equal to the sound-pressure level in decibels, relative to 0.0002 microbar, of a 1000-hertz tone that is considered by listeners to be of equal loudness with the sound under consideration.

phone—*See* Headphone.

phone jack—Also called a telephone jack. A jack designed for use with phone plugs.

phone plug—Also called telephone plug. A plug used for quick connection of headphones, microphones, and other audio equipment.

phonograph—An instrument for reproducing sound. It consists of a turntable on which the grooved medium containing the impressed sound is placed, a stylus that rides in the groove, and an electrical (formerly mechanical) amplifying system for converting minute vibrations of

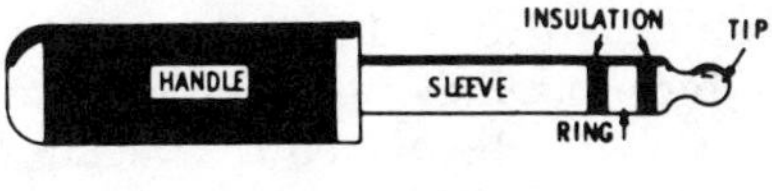

Phone plug.

the needle into electrical impulses that drive a speaker.

phonograph oscillator—An rf oscillator circuit, the output of which is modulated by a phonograph pickup and sent through space to a receiver. Thus, no wires to the receiver are needed.

phonograph pickup—A transducer that is actuated by modulations in the groove of a recording medium and transforms this mechanical input into an electric output.

phono jack—A jack designed to accept a phono plug.

phono pickup—*See* Phonograph Pickup.

phono plug—A small plug used with a phono jack for making connections between a phono cartridge and a preamplifier or between preamplifier and amplifier, etc. The plug is usually mounted on the end of a flexible, shielded cable.

photoconductive cell—A photoelectric cell, having a resistance that varies with the intensity of illumination striking it.

photoelectric cell—Also called photocell. A cell, such as a photovoltaic or photoconductive cell, whose electrical properties are affected by illumination. The term is sometimes misapplied to a phototube, which is a vacuum tube and not a cell.

photoelectric conductivity—The increased conductivity exhibited by certain crystals when exposed to light (e.g., a selenium cell).

photoelectric effect—The effect that incident light has on certain materials, increasing the number of free electrons available for electric current.

photoelectric phonograph pickup—A phonograph reproducing device consisting essentially of a light source, a stylus assembly to which a very thin mirror is attached, and a selenium cell that picks up light reflected by the mirror. Sidewise movements of the stylus in the record groove cause the amount of reflected light to vary and this varies the resistance of the selenium cell. Accordingly, an electrical signal is developed that corresponds to the audio in the modulated groove of the record.

photoelectric transducer—A transducer that converts changes in light intensity into variations in electric potential.

photophone—A device for converting variations in light intensity into sound.

photosensitive recording—Recording by the exposure of a photosensitive surface to a signal-controlled light beam or spot.

photovoltaic cell—A cell made of semiconductor material that produces many free electrons when exposed to light. These electrons are available for electric current without the necessity of applying an external potential.

photovoltaic effect—The generation of a voltage (or an electric field) in a material that is illuminated with radiation of a suitable wavelength.

pi—Greek letter symbolized by π. It designates the value of the ratio of the circumference of a circle to the diameter, or approximately 3.14159.

pickup—1. A device that converts a sound, scene, or other form of intelligence into corresponding electric signals (e.g., a microphone, television camera, or phonograph pickup). 2. Interference from a nearby circuit or system.

pickup arm—*See* Tone Arm.

pickup cartridge—The removable assembly at the end of a phonograph tone arm that contains the transducer elements and the stylus.

Pickup cartridge (Shure Brothers Inc. Model V15 Type II).

pico—Prefix meaning 10^{-12}. (Formerly micromicro.) Abbreviated p.

picoampere—One-millionth of a microampere (10^{-12} ampere). Abbreviated pA.

picofarad—One-millionth of a microfarad (10^{-12} farad). Abbreviated pF. Formerly micromicrofarad.

picosecond—One-millionth of a microsecond (10^{-12} second). Abbreviated ps.

picowatt—One-millionth of a microwatt (10^{-12} watt). Abbreviated pW.

piezodielectric—A property of certain materials that causes a change in their dielectric constant under mechanical stress.

piezoelectric—A property of certain crystals that causes them to produce a voltage when subjected to mechanical stress or, conversely, to exhibit mechanical stress when subjected to a voltage.

piezoelectric effect—1. The mechanical deformity of certain natural and synthetic crystals under the influence of an electric field. This effect is used in high-precision oscillators and certain high-frequency filters. 2. The property of certain natural and synthetic crystals to produce a voltage when subjected to mechanical stress (compression, expansion, twisting, etc.).

piezoelectric microphone— *See* Crystal Microphone.

piezoelectric pickup—*See* Crystal Pickup.

piezoelectric speaker—*See* Crystal Speaker.

piezoelectric transducer— Also called ceramic or crystal transducer. A transducer that depends for its operation on the piezoelectric properties of certain materials: the production of an electric charge when stressed and, conversely, the production of mechanical movement when an electrical signal is applied.

pilot subcarrier—A subcarrier used as a control signal in the reception of compatible fm stereophonic broadcasts.

pinch effect—In the reproduction of lateral recordings, the pinching of the reproducing stylus tip at two points in cycle. This results because the wedge-shaped cutting stylus cuts a narrower groove at the middle of its excursions than it does at the ends.

pinch roller—*See* Pressure Roller.

ping-pong effect—The apparent shifting of a sound source back and forth between the two sides of a stereophonic reproducing system.

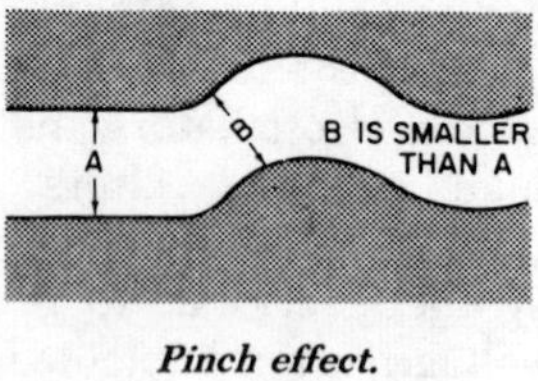

Pinch effect.

pink noise—Noise with a spectral intensity that is inversely proportional to frequency over a specified range.

piston action—The movement of a speaker cone or diaphragm when driven at the bass audio frequencies. The entire diaphragm moves as a unit rather than flexing in subunits of area.

pitch—1. That attribute of auditory sensation by which a listener judges a tone as being higher or lower on a musical scale. 2. The distance between two successive unmodulated grooves of a disc recording.

piv—Abbreviation for peak inverse voltage.

plate—1. Preferably called anode. The principal electrode to which the electron stream is attracted in a vacuum tube. 2. One of the conductive electrodes in a capacitor. 3. One of the electrodes in a storage battery.

platter—A popular term for phonograph records and transcriptions.

playback—The process by which a recording is made to yield the original recorded sound.

playback head—A magnetic head used to pick up a signal from a magnetic tape.

playback loss—*See* Translation Loss.

plug—The movable half of the two mating halves of a connector. The plug is usually thought of as the male portion of the connector, but this is not always true.

plug connector—An electrical fitting containing male, female, or male and female contacts and constructed so that it can be attached to the end of a wire, cable, etc., for joining with another electrical connector or connectors.

pm—Abbreviation for phase modulation or permanent magnet.

pm erase head—A head that uses the fields of one or more permanent magnets for erasing.

pm speaker—Abbreviation for permanent-magnet speaker.

PndB—Perceived noise level expressed in decibels. *See* Perceived Noise Level.

pneumatic speaker—A speaker in which the acoustic output is produced by modulating an air stream with the audio signal.

pnp transistor—A transistor consisting of two p-type regions separated by an n-type region.

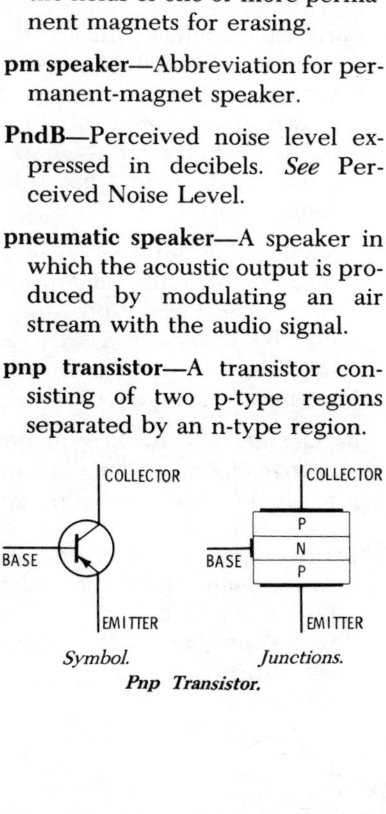

Symbol. *Junctions.*
Pnp Transistor.

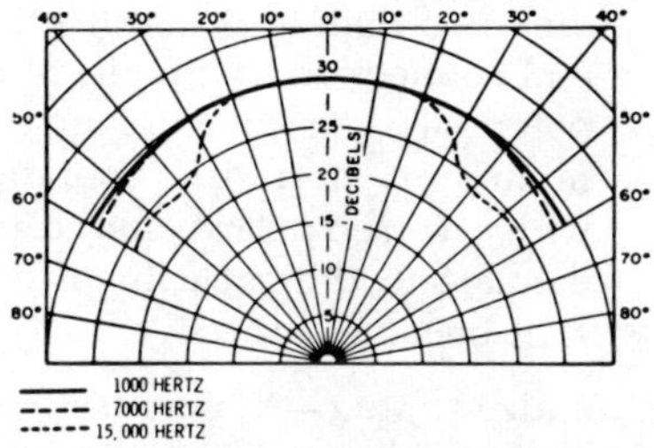

Polar diagram (response of a speaker enclosure).

point source—A radiation source whose dimensions are small compared with the distance from which it is observed.

polar diagram—A diagram using polar coordinates, on which it is possible to plot vector quantities—that is, quantities that have both direction and magnitude. Often used to plot sound-radiation patterns.

polar grid—A type of circular grid on which range and azimuth are represented from a central reference point.

polarization of a dielectric—The slight displacement of positive and negative charges in the molecules of a dielectric material when it is placed in an electric field.

polarized—1. Having two opposite electric charges, one positive and the other negative. 2. Having two opposite magnetic poles, one north and the other south.

polarized capacitor—An electrolytic capacitor with the dielectric film formed adjacent to only one metal electrode. Direct current then passes more freely in one direction than in the other. Normal connection to circuits of dc potential is in the direction of minimum current; operation with reversed polarity may cause damage to the capacitor if current becomes excessive.

polarized plug—A plug so constructed that it may be inserted in its receptacle only in a predetermined position.

polarized receptacle—A receptacle into which a polarized plug can be inserted only in a predetermined position.

polar radiation pattern—A diagram that shows the relative strength of radiation from a source in all directions in a given plane.

pole—1. One end of a magnet. 2. One electrode of a battery (negative or positive). 3. An output terminal on a switch.

pole piece—One or more pieces of ferromagnetic material forming one end of a magnet and so shaped that the magnetic flux is concentrated or directed in a desired pattern.

polyester—Polyethylene glycol terephthalate, a material often used as a base film for precision magnetic tape. Characterized by high stability with respect to humidity and time, resistance to solvents, and mechanical strength.

polyester backing—A plastic film backing added to magnetic tape to make it stronger and more resistant to changes in humidity.

polyethylene—Short for polymerized ethylene, a tough, white plastic insulator with low moisture absorption. It is often used as a dielectric.

polystyrene—A clear thermoplastic material having excellent dielectric properties, especially at ultrahigh frequencies.

polystyrene capacitor—A low-loss precision capacitor with a polystyrene dielectric.

polyvinyl chloride—A thermoplastic widely used for insulations and jackets on components, wire, and cable.

port—An opening giving access to a system or circuit. It allows energy to be supplied or withdrawn or measurements to be made. Examples are the port in a waveguide or in a bass-reflex speaker enclosure.

portable recorder—A sound recorder of a size and weight that make it easily portable. It may be self-powered (batteries) or require external connection to a power line.

portamento— The continuous change of a tone from one pitch to another. (*See also* Glissando.)

positive—A term chosen arbitrarily to represent electric potentials, power terminals, battery electrodes, etc., to which electrons are attracted.

positive charge—An electrical charge with fewer electrons than normal.

positive electrode—The element of a primary cell that connects to the positive terminal of the cell and receives electrons from any external circuit connected to it.

positive feedback—The process whereby part of the output power is returned to the input in proper phase to reinforce the input power. (*See also* Regeneration.)

positive-going—Increasing in a positive direction (said of a current or waveform).

positive ion—An atom that has lost one or more electrons and thus has an excess of protons, giving it a positive charge.

positive magnetostriction—Magnetostriction in which a material expands whenever a magnetic field is applied.

positive temperature coefficient—The condition whereby some attribute of a material or component such as resistance, capacitance, or length increases with an increase in temperature.

positive terminal—In a battery or other voltage source, the terminal toward which electrons flow through the external circuit from the negative terminal.

post-emphasis—*See* De-emphasis.

post-equalization—*See* De-emphasis.

potential difference—The potential difference between two points in an electric circuit is equal to the work, because of electrostatic forces, involved in the transfer of a charge between the two points, divided by the magnitude of the charge. Expressed as a formula:

$$V = \frac{W}{Q}$$

where,
V is potential difference in volts,
W is work in joules,
Q is charge in coulombs.

potential divider—*See* Voltage Divider.

potential drop—The difference in electric potential between the two ends of a current-carrying resistance.

potentiometer—1. An electronic component consisting of a resistive element with terminals for connecting to each end and a third terminal connecting to a wiper contact with adjustable range over all or part of the resistance element. 2. A unit for measuring potential difference, using a calibrated voltage divider, a reference potential, and a null detector.

powdered-iron core—A core consisting of fine particles of magnetic material mixed with a suitable bonding material and molded to shape.

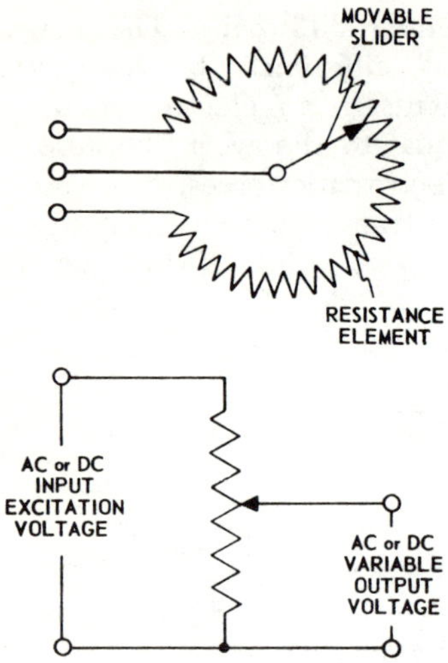

Potentiometer, 1.

power—The rate at which work is done. Units of electrical power are the watt and the kilowatt.

power amplifier—An amplifier used to boost audio signals to sufficient strength for speaker operation.

power cord—*See* Line Cord.

power dissipation—The dispersion of heat generated by current through the resistance of an electrical device. The heat is dissipated by radiation, conduction, or convection.

power factor—1. Ratio of the actual, wattmeter power of an alternating current to the apparent power, as indicated by an ammeter and voltmeter. 2. Ratio of resistance to impedance in a complex circuit—therefore, a measure of the loss in an inductor, capacitor, or insulator. 3. The cosine of the phase angle between voltage applied to a load and current through it.

power-factor correction— The practice of adding capacity to an inductive circuit to increase the power factor by making the total current more nearly in phase with the applied voltage.

power frequency—The frequency at which electric power is supplied by the electric power company. Throughout most of the United States, the frequency is 60 Hz.

power gain—Also called power amplification. The ratio of power delivered to a specified load impedance by an amplifier to the power absorbed by the amplifier input.

power line—Two or more wires conducting electric power from one location to another.

power output—The power in watts delivered by a power amplifier to a load such as a speaker.

power rating—The maximum power that can be dissipated in a component or device for a given length of time.

power ratio—Ratio of the power output to the power input of a device. Usually expressed as the number of decibels loss or gain.

power response—The frequency-response capabilities of an amplifier running at or near its full rated power.

ppm—1. Abbreviation for pulse-position modulation. 2. Abbreviation for parts per million.

pps—Abbreviation for pulses per second.

practical system of electrical units—A system in which the units are multiples or submultiples of the units of the centimeter-gram-second electromagnetic system.

praetersonic—The higher region of the sonic spectrum.

praetersonics—The propagation and signal processing of acoustic waves in solids at frequencies that extend into the microwave region.

preamplifier—An amplifier used primarily to raise the output of a low-level source to a usable level for a power amplifier. A preamplifier often includes provision for equalization and mixing.

predistortion—*See* Pre-emphasis.

pre-emphasis—Also called pre-equalization or predistortion. In recording, an arbitrary change in the frequency response from its basic response (e.g., constant velocity or amplitude) to improve the signal-to-noise ratio or to reduce distortion.

pre-emphasis network—A network inserted into a system to emphasize one range of frequencies.

pre-equalization— *See* Pre-emphasis.

prefix multiplier—Prefixes that designate a greater or smaller unit than the original, by the factor indicated. The following are some prefixes commonly used:

Prefix	*Symbol*	*Factor*
tera	T	10^{12}
giga	G	10^{9}
mega	M	10^{6}
kilo	k	10^{3}
hecto	h	10^{2}
deka	da	10
deci	d	10^{-1}
centi	c	10^{-2}
milli	m	10^{-3}
micro	μ	10^{-6}
nano	n	10^{-9}
pico	p	10^{-12}

preform—Also called a biscuit. In disc recording, the small slab of record material used to press a record.

prerecorded tape—*See* Recorded Tape.

presence—The quality of naturalness in sound reproduction. When the presence of a system is good, the listener gains the impression that the performers and their instruments are actually present in the listening room.

presence control—A potentiometer used in a three-way speaker system for controlling the volume of the middle-range speaker.

pressing—A disc recording produced in a record-molding press from a master or stamper.

press-to-talk switch—Also called a push-to-talk switch. A spring-loaded microphone switch that must be held down as long as the operator talks. Releasing the switch deactivates the microphone. It is used on transmitter and dictating-machine microphones.

pressure—Force per unit area. Measured in pounds per square inch (psi) or by the height (in feet, inches, or centimeters) of a column of water or mercury that the force will support. Absolute pressure is measured with respect to zero pressure. Gauge pressure is measured with respect to atmospheric pressure.

pressure microphone—A microphone with output that corresponds to the instantaneous sound pressure of the impressed sound waves. It is nondirectional when its dimensions are smaller than a wavelength of the intercepted sound.

pressure pads—In tape recorders, small pieces of felt or similar soft material that hold the tape in intimate contact with the heads during record or playback.

pressure roller—Also called pinch roller, puck, or capstan idler. A spring-loaded, rubber-tired roller that presses the magnetic tape tightly against the capstan. As the capstan revolves, it pulls the tape past the record or playback heads.

pressure transducer—A device that responds to inputs of static or dynamic pressure and produces a proportional electric output.

prf—Abbreviation for pulse-repetition frequency.

primary cell—A cell that produces electric current by means of an electrochemical reaction that is not reversible. Therefore, the cell cannot be recharged; once it is discharged it must be discarded.

primary circuit—The first, in electrical order, of two or more coupled circuits, wherein a change in current will induce a voltage in the other, or secondary, circuit.

primary current—The current flowing through the primary winding of a transformer. Changes in this current cause a voltage to be induced in the secondary winding of the transformer.

primary standard—A standard of higher accuracy than a secondary standard and against which all secondary standards of that type are calibrated.

print-through—The transfer of magnetism from one layer of tape to another layer while the tape reel is stored.

propagation velocity—*See* Velocity of Propagation.

prr—Abbreviation for pulse-repetition rate.

ps—Abbreviation for picosecond (10^{-12} second).

pseudostereo—Equipment and methods for obtaining stereo qualities from one channel of audio reproduction.

ptm—Abbreviation for pulse-time modulation.

p-type—Semiconductor material that has been doped with an excess of acceptor impurity atoms, so that free holes are available to conduct electricity.

public-address system—Also called a pa system. A system for picking up and amplifying sounds to be delivered to a large audience. The system may consist of one or more microphones, mixers, amplifiers, and speakers.

puck—*See* Pressure Roller.

pulsating current—A nonuniform electron flow that varies periodically but does not reverse its direction.

pulse - amplitude modulation—Abbreviated pam. Modulation in which the modulating wave is caused to amplitude-modulate a pulse carrier.

pulse - duration modulation—Abbreviated pdm. Pulse-time modulation in which the duration of a pulse is varied. Also called "pulse-width modulation" and "pulse-length modulation."

pulse-interval modulation—Pulse-time modulation in which the pulse spacing is varied.

pulse modulation—The use of a series of pulses modulated to convey information. Modulation may involve changes of pulse amplitude (pam), position (ppm), or duration (pdm).

pulse - position modulation—Abbreviated ppm. Pulse modulation in which the value of each instantaneous sample of the modulating wave controls the position in time of a pulse.

pulse rate—*See* Pulse-Repetition Rate.

pulse ratio—Ratio of the length of any pulse to its total period.

pulse repetition rate—Abbreviated prr. Also called pulse rate. The average number of pulses per unit of time.

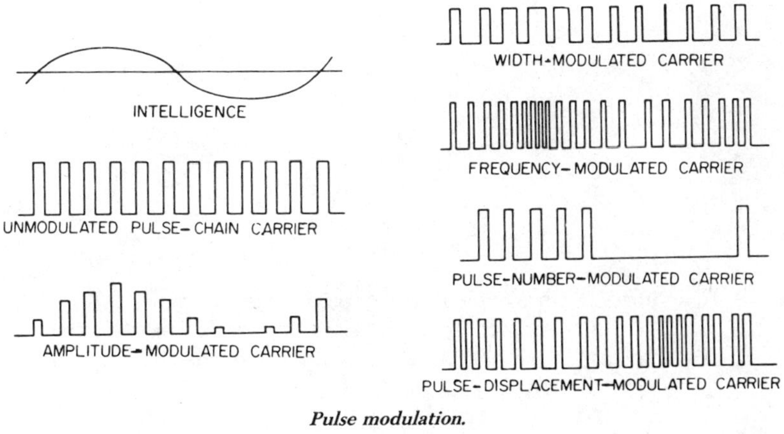

Pulse modulation.

pulse rise time—The interval of time required for the leading edge of a pulse to rise from 10% to 90% of its peak amplitude, unless some other percentage is stated.

pulse spacing modulation—Formerly called pulse-interval modulation. A type of pulse-time modulation that varies the pulse spacing.

pulse time—The time interval measured from a point at 90% of maximum amplitude on the leading edge of a pulse, to the 90% point on the trailing edge.

pulse-width modulation—*See* Pulse-Duration Modulation.

pure tone—*See* Simple Tone.

push-pull amplifier—*See* Balanced Amplifier.

push-pull circuit—A circuit containing two similar active elements that operate in 180° phase relationship. The input to one element is negative going while the input to the other is positive going. The outputs are added in series. Cancellation of certain unwanted products is achieved, plus increased power output.

push-pull microphone—A microphone combining two like elements actuated by the same sound waves and operating 180° out of phase.

push-to-talk switch—*See* Press-to-Talk Switch.

pW—Abbreviation for picowatt.

Pythagorean scale—A musical scale in which the frequency intervals are regulated by mathematical ratios rather than by consonances.

Q

Q—1. In an inductor or a capacitor, the ratio of its reactance to its effective series resistance at a given frequency. Also called quality factor or *Q* factor. 2. A measure of the sharpness of resonance or frequency selectivity of a mechanical or electrical system. 3. Symbol for quantity of electric charge.

Q-8—Name used by RCA for their four-channel, eight-track tape cartridges.

***Q*-factor**—*See Q*, 1.

QS—A matrixing system for encoding four-channel sound into two channels. Developed by Sansui Corporation of Japan.

quadradisc—Another name for the CD-4 discrete recording system developed by RCA/JVC.

quadraphonic sound—The four-channel stereo sound system that uses four separate signal paths from microphones to speakers (discrete) or one of several matrixing systems to carry the same information on only two channels.

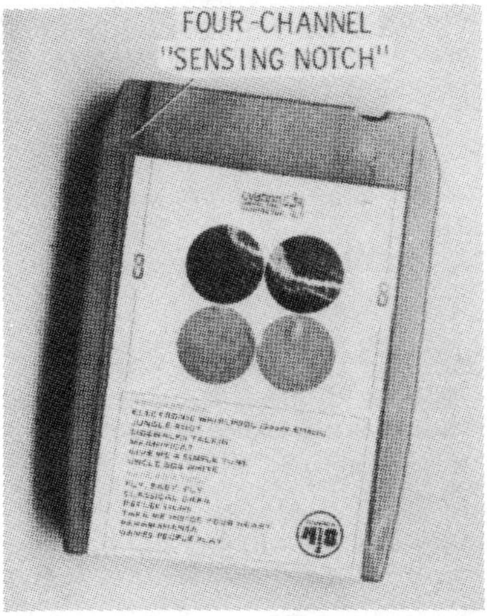

Q-8 tape cartridge.

quadraphonic tape track layout—The spacing arrangement on a recorded tape that permits quadraphonic program material to be recorded and played back from a single tape. (*See* illustration.)

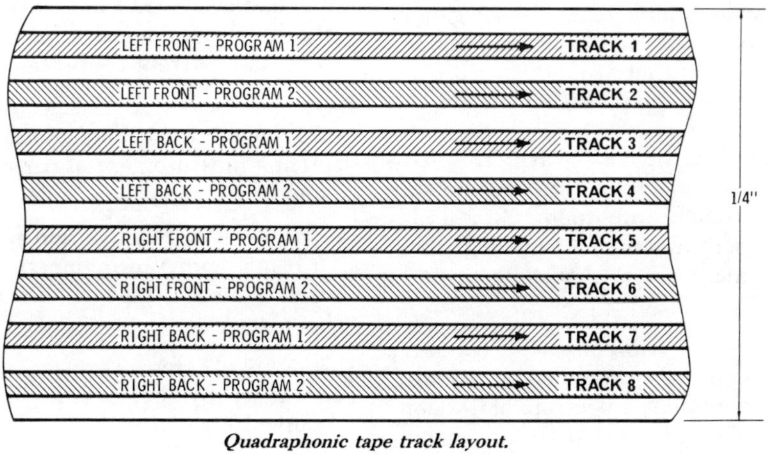

Quadraphonic tape track layout.

R

R—Symbol for resistance, resistor, or reluctance.

radian—In a circle, the angle subtended by an arc equal to the radius of the circle. Numerically, it is equal to 57° 17′ 44.8″. A complete circle contains 2π radians. One degree equals 0.01745 radian.

radian length—The distance, in a sinusoidal wave, between phases differing by an angle of 1 radian. It is equal to the wavelength divided by 2π.

radian per second—A unit of angular velocity.

random-noise generator—A generator of a succession of random signals that are distributed over a wide frequency spectrum.

random signals—Waveforms having at least one parameter (usually amplitude) that is a random function of time (e.g., thermal noise or shot noise).

random vibration—A vibration generally composed of a broad, continuous spectrum of frequencies, the instantaneous magnitude of which cannot be specified to any given moment of time. (*See also* White Noise.)

rated output—The output power, voltage, current, etc., at which a machine, device, or apparatus is stated as capable of operating without deleterious effects.

rate limiting—Nonlinear behavior in an amplifier that is unable to respond to large, rapid changes in output voltage.

rate of decay—The rate at which the sound-pressure level is decreasing at a given point at a given time. The practical unit is the decibel per second.

ratio of transformation—The ratio of the secondary voltage of a transformer to the primary voltage under no-load conditions, or the corresponding ratio of currents in a current transformer.

Rayleigh disc—A special acoustic radiometer used for the fundamental measurement of particle velocity.

raw tape—Also called virgin tape or blank tape. A term sometimes used to describe tape that has not been recorded.

RC—Symbol for resistance-capacitance. Symbol for resistance-coupled.

RC amplifier—Abbreviation for resistance-capacitance coupled amplifier.

RC circuit—A time-determining network of resistors and capacitors in which the time constant is defined as resistance times capacitance.

RC constant—The time constant of a resistor-capacitor circuit, equal in seconds to the resistance in ohms multiplied by the capacitance in farads.

RC coupling—*See* Resistance-Capacitance Coupling.

RC filter—*See* Resistance-Capacitance Filter.

RC network—A circuit containing resistances and capacitances arranged in a configuration to produce a desired effect such as filtering, frequency discrimination, and control of charge and discharge rates.

RC oscillator—An oscillator in which the frequency is determined by resistance-capacitance elements. (*See also* Resistance-Capacitance Oscillator.)

reactance—Symbolized by X. Opposition to the flow of alternating current. Capacitive reactance (X_C) is the opposition offered by capacitors, and inductive reactance (X_L) is the opposition offered by a coil or other inductance. Both reactances are measured in ohms.

reactive load—A load having a net reactance (i.e., a capacitive or inductive load) at the frequency of operation, as opposed to a resistive load.

reactive power—Also called wattless power. In an ac circuit, the product of volts times amperes, when they are 90° out of phase. The unit of measurement is the var (stands for volt-amperes reactive).

real power—The component of apparent power that represents true work in an ac circuit. It is expressed in watts and is equal to the apparent power times the power factor.

receptacle—Usually the fixed or stationary half of a two-piece multiple-contact connector. Also, the connector half usually mounted on a panel and containing socket contacts.

record changer—A device that will automatically play a number of phonograph records in succession.

record compensator—Also called a record equalizer. An electrical network that compensates for different frequency-response curves in various recording techniques.

recorded tape—Also called a prerecorded tape. Tape that may be purchased already containing recorded program material.

record equalizer—*See* Record Compensator.

recorder—An instrument that makes a permanent record of varying electrical impulses, that, in turn, are derived from various forms of intelligence such as codes, sound (music, voices, etc.), and visual programs (tv or live performances). Some of the recording mediums used are disc, optical film, and magnetic tape or wire.

recording blank—*See* Recording Disc.

recording channel—One of several independent recorders in a recording system, or independent recording tracks on a recording medium.

recording disc—Also called a recording blank. A blank (unrecorded) disc made for recording purposes.

recording head—A magnetic head that transforms electrical variations into magnetic variations for storage on magnetic media. (*See also* Cutter.)

recording level—The amplifier output required to provide a satisfactory recording.

recording loss—In disc recording, the loss in recorded level that occurs because the amplitude of recording stylus deflection differs from the amplitude of the wave in the recording medium.

Recording head.

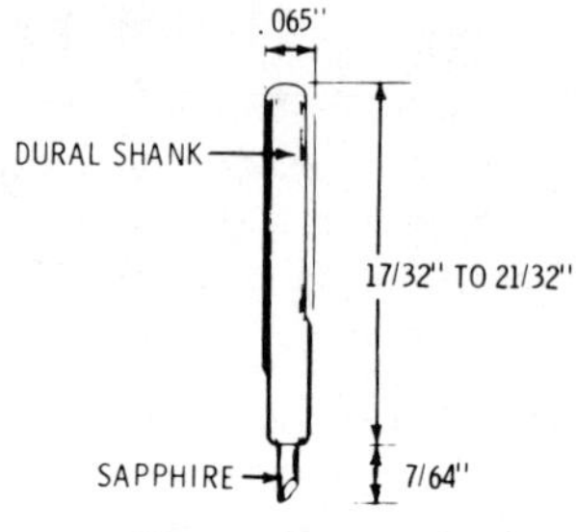

Recording stylus.

recording noise—Unwanted audio signals induced on the tape by the preamplifier, amplifier, or other components of a recording system.

recording stylus—The tool that inscribes a groove in the recording medium.

record player—An assembly consisting of a motor-driven turntable, pickup arm, and pickup cartridge, for converting the undulations of the record grooves into corresponding audio-frequency voltages. These voltages are then applied to an amplifier/speaker system for conversion to sound waves.

rectification—The conversion of alternating current into unidirectional or direct current.

rectifier—A device that will conduct easily in one direction but with difficulty in the other direction. Because of this unidirectional characteristic it is often used to convert alternating current to direct current.

reference acoustic pressure—The magnitude of a complex sound that produces a sound-level meter reading equal to the reading that results from a sound pressure of 0.0002 dyne per square centimeter at 1000 hertz. Also called reference sound level.

reference level—The zero or starting point for measuring the value in dB of an alternating quantity. For sound loudness, the reference level is usually the threshold of hearing. A common reference in electronics is one milliwatt, and power is usually stated as so many decibels above or below this figure.

reference sound level—*See* Reference Acoustic Pressure.

reflected impedance—1. The apparent impedance across the primary of a transformer when current flows through a load across the secondary. 2. The impedance at the input terminals of a transducer as a result of the impedance characteristics at the output terminals.

reflected power or signal—The power flowing back to the generator from the load.

reflected resistance—The apparent resistance across the primary of a transformer when a resistive load is placed across the secondary.

reflex baffle—A speaker baffle in which a portion of the radiation from the rear of the diaphragm is subjected to a phase shift and used to reinforce certain relatively weak frequencies radiated at the front of the diaphragm.

refraction—The change in direction of propagation of a wavefront because of its passing from one medium into another in which its speed is different. Refraction may also occur in a single medium of varying characteristics.

regeneration—The gain in power that occurs when part of the output of an amplifier is fed back to the input in a phase that increases the input signal. Also called regenerative feedback and positive feedback.

register—A range of notes used for playing a particular piece or a part of it (e.g., melody or harmony), particularly the range covered by a clavier or manual.

regulated power supply—A power unit that tends to maintain a constant output voltage or current in spite of variations in line voltage, current demand, load impedance, etc.

rel—A unit of reluctance, equal to one ampere-turn per magnetic line of force.

relative permeability— Ratio of magnetic permeability of one material to that of another, or of the same material under different conditions.

relative power—A power level referred to another power level.

relative response—The ratio, usually expressed in decibels, of the response under some particular conditions to the response under a stated reference condition.

reluctance—The opposition of a magnetic path to the flow of magnetic lines of force through it. The reciprocal of permeance, it equals the magnetomotive force divided by the magnetic flux.

remanence—Another term for retentivity. The extent to which a body remains magnetized after a magnetizing force that has brought the body to saturation returns to zero.

remanent magnetization— The magnetization retained by a substance after the magnetizing force has been removed.

repetition frequency—The number of repetitions of an event per unit time.

reproduce head—An electromagnetic transducer used to play back a recorded tape.

reproducer—A device used to translate electrical signals into sound waves.

reproducing stylus—The flexible part of a phonograph pickup that engages the grooves of a record and transmits the groove modulation to the transducer element.

repulsion-induction motor—A single-phase motor in which the stator winding is connected to the source of power and the rotor winding is connected to the commutator. Brushes on the commutator are short-circuited and are placed so that the magnetic axis of the rotor winding is inclined to that of the stator winding. This type of motor has a high starting torque. When the armature attains operating speed, a centrifugal device short-circuits the entire armature and the motor then runs as a single-phase induction motor.

rerecording—The process of making a recording by reproducing a recorded sound source and recording this reproduction. (*See also* Dubbing.)

rerecording system—A collection of recorders, mixers, amplifiers, etc., that can be used to combine program material of varied content (music, speech, and sound effects) in a single recording.

residual field—The magnetic field that remains in a magnetic material after the original magnetizing force has been removed.

residual flux—In a magnetic circuit, the value of magnetic induction remaining after the magnetomotive force has been removed.

residual flux density—The magnetic flux density that remains in a magnetic material that has been cyclically magnetized and has had the magnetizing field reduced to zero.

residual induction—Magnetic induction remaining in a magnetized material after the effective magnetizing force has been reduced to zero.

residual losses—Any losses in the core of a magnetic circuit, other than the hysteresis and eddy current losses.

residual magnetic induction—In a magnetic material, the magnetic induction that remains after the magnetizing force has been removed.

residual magnetism—The magnetism remaining in a magnetic material after the magnetizing force has been removed.

resistance—A property of conductor materials that opposes the passage of steady direct current. The unit of resistance is the ohm, which is defined as the resistance that allows 1 ampere to flow when 1 volt of potential difference is applied. This relation is expressed by Ohm's law, $R = E/I$, where R is a constant. This concept of resistance is often extended to cases where R is not a constant but changes with different values of E. R is then called a nonlinear resistance. In an electrical circuit, resistance dissipates power in the form of heat.

resistance-capacitance-coupled amplifier—An amplifier having stages interconnected by a combination of resistors and capacitors.

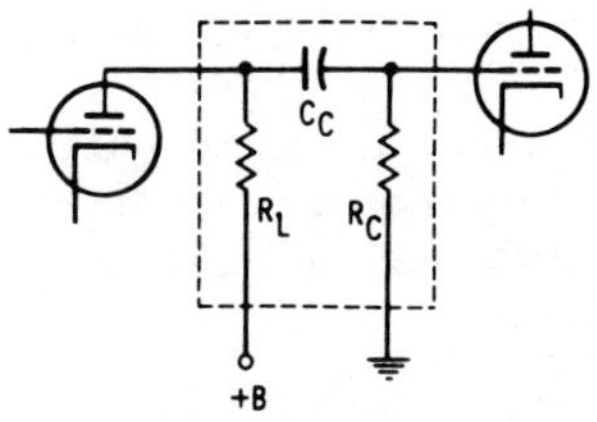

Resistance-capacitance coupling.

resistance-capacitance coupling—Also called RC coupling. Coupling between two or more circuits by a combination of resistive and capacitive elements.

resistance-capacitance filter—A filter having for its components only resistors and capacitors. Abbreviated RC filter.

resistance drop—Commonly called IR drop. The drop in circuit potential that occurs across a resistance because of current through it. The drop in volts is equal to current (I) in amperes times resistance (R) in ohms.

resistance loss—Power lost in an electrical circuit because of current through resistance. Measured in watts, its value is equal to the current squared times the resistance ($W = I^2R$).

resistance strain gauge—A gauge or transducer constructed of material that undergoes a change in resistance when its dimensions are subjected to a stress, resulting in an elastic displacement, or strain. This principle has been used in the design of certain phonograph pickups.

resistive load—A load having only resistance, or if reactance is present, capacitive and inductive reactances are equal and cancel at the frequency involved, so that voltage and current are in phase.

resistivity—Symbolized by the Greek letter rho (ρ).

Resistivity,

$$\rho = \frac{RA}{l}$$

where,

R is total resistance of a conductor, in ohms,

A is area of conductor in square meters,

l is length of conductor in meters.

Hence the unit of resistivity, ρ, is in ohm-meters. Since the resistivity of copper is about 1.673 X 10^{-8} times this unit, a smaller unit is sometimes used, the *ohm per mil foot.* A in the formula is stated in circular mils (the area of a circle 0.001 inch in diameter) and *l* is in feet. Using this unit, the resistivity of copper is about 10.06.

resolution—1. The analyzing or resolving of a sound into its smallest individual components so that it may be later synthesized at will, either wholly or in part. 2. In describing recording tape, the smallest magnetic domains that can be recorded and reproduced in playback. This factor, with head-gap size and tape speed, determines the upper limit of frequency response.

resonance—1. In a circuit containing capacitive and inductive reactance, the condition where these two reactances are equal and cancel each other. 2. In a vibrating sound system, with the vibrating member being a stretched string, elastic reed or diaphragm, air column, etc., the condition in which the natural vibration period of the member is in step with the exciting impulse and reinforces it.

resonance curve—A graph of the response of a resonant system showing how the output of the system varies for frequencies at or near the resonant frequency.

resonance indicator—A device that produces some indication, with meter, neon lamps, vibrating reeds, headphones, etc., that a system is at resonance.

resonant circuit—A circuit containing inductance and capacitance and that is therefore resonant to a certain frequency, depending on the values of such inductance

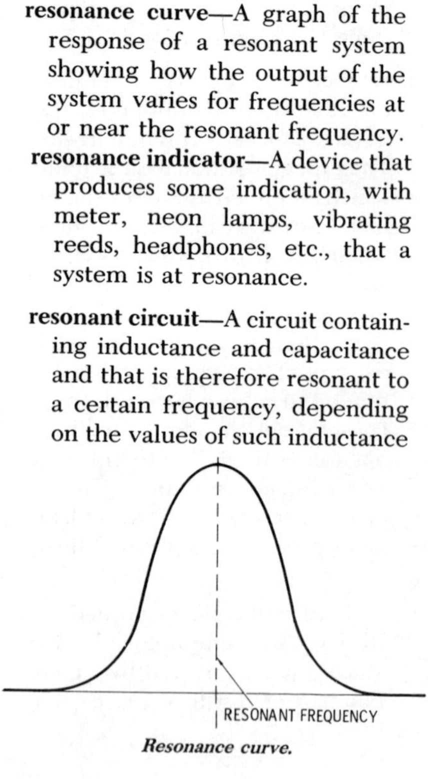

Resonance curve.

and capacitance. If either or both of these components are adjustable, the circuit is said to be tuned.

resonant frequency—1. In an LC circuit, the frequency at which the inductive reactance equals the capacitive reactance, the impedance is minimum, or the current is in phase with the applied voltage. The formula for resonant frequency is:

$$f = \frac{1}{2\pi\sqrt{LC}}$$

where,

f is the resonant frequency in hertz,
L is inductance in henrys,
C is capacitance in farads.

2. In a vibrating system, the frequency of applied pulses that are in step with the natural vibration period of the system.

resonator—A system or device that reinforces or sustains the oscillations of another system. (*See also* Helmholtz Resonator.)

response curve—A graph plotting amplitude of output against frequency of input for an amplifier, speaker, pickup, sound system, etc.

retentivity—The ability of a material, once magnetized, to retain magnetism after the magnetizing force is removed. As an example, soft iron is easily magnetized, but loses magnetism easily when the force is removed. Therefore, it is said to have a low retentivity.

reverberation—An increase in sound duration beyond the duration of the original sound impulse because of reflections from walls, floors, ceilings, etc., in the vicinity of the sound source.

reverberation chamber—An enclosure with sound-reflecting surfaces designed to prolong sound. It is used as a testing room for certain acoustic measurements.

reverberation time—The time required for a sound to decay to one millionth (–60 dB) of its original intensity.

reverberation unit—A device used to impart a sense of liveness to reproduced or transmitted sound by artificial echoes or reflections.

rewind—In reel-to-reel recorders, the function of returning the tape to its starting position.

rhythm bar—A bar used in some electronic organs to provide a rhythmic chord accompaniment as long as the bar is depressed.

RIAA curve—A recording characteristic curve recommended as standard for long-playing records by the Record Industry Association of America.

ribbon microphone— A microphone having the moving and signal-generating element in the form of a very light weight metallic ribbon. Characterized by wide frequency range, low distortion, and low internal noise.

right (or left) stereophonic channel—The section of a stereophonic amplifier that receives the signal recorded from the right (or left) microphone of a stereo recording system and applies the amplified signal to the right (or left) speaker in the playback system.

rim drive—A method of driving a phonograph turntable by using a rotating, rubber-covered wheel that contacts the inner surface of the rim of the turntable.

ring head—A magnetic recording head having the magnetic material shaped more or less in the form of a ring, with one or more air gaps to direct the magnetic flux through the recording medium.

ringing—High-frequency damped oscillations typical of an audio system that is subjected to regenerative audio feedback.

ring magnet—A ceramic permanent magnet in the shape of a ring; this imposes certain dimensional restrictions to maintain the ring shape. For example, the axial length should be no greater than the wall thickness, which, in turn, should be no less than 15% of the outside diameter.

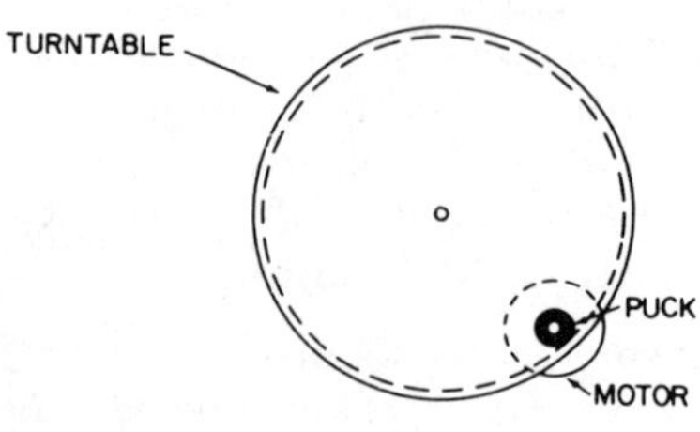

Rim drive (turntable).

ripple current—The alternating component of an otherwise steady direct current. A term usually used in describing how far the output of a power supply departs from ideal dc.

ripple filter—A filter for removing the ripple component from the output of a dc supply. Also called smoothing filter.

ripple frequency—Frequency of the ripple component of a dc voltage. In full-wave rectifier supplies, it is twice the frequency of the line voltage.

ripple voltage—The ac component in the output of a power supply or other dc source. It is usually a small percentage of the total voltage, in well-designed supplies.

rise time—A measure of the steepness of the leading edge of a waveform. It is the time required for the pulse to rise from 10% to 90% of its final value.

RMA—Abbreviation for Radio Manufacturers Association. Now EIA (Electronics Industries Association).

rms—Abbreviation for root-mean-square.

rms amplitude—The root-mean-square amplitude of a cyclically repeating waveform such as an ac voltage or current. It is equal to the square root of the average of the squares of all instantaneous values of the waveform over a period of one cycle. Also called effective value because it will produce the same effective power dissipation across a given resistance as a dc voltage or current of the same value. For a sine wave, the rms value is equal to 0.707 times the peak value.

rms voltage—The effective or root-mean-square value of an alternating voltage. The value that will produce the same heating effect or power dissipation on a pure resistance as a dc voltage of the same value. For a sine wave, the rms voltage is equal to 0.707 times the peak voltage.

Rochelle-salt crystal—The crystal form of the chemical compound sodium potassium tartrate. Because of its piezoelectric properties it finds use in microphones, phonograph pickups, and record-cutting heads.

rolloff—A term used to describe the action of certain audio filters designed to attenuate frequency response at a predetermined rate beyond a chosen point.

room noise—*See* Ambient Noise.

root-mean-square— Abbreviated rms. The square root of the mean, or average, of all the squares of the instantaneous values of a cyclically repeating quantity (voltage, current, etc.). Also called effective value because it represents the effective power of the quantity.

root-mean-square amplitude—*See* Rms Amplitude.

root-mean-square voltage— *See* Rms Voltage.

rpm—Abbreviation for revolutions per minute.

rps—Abbreviation for revolutions per second.

RTMA—Abbreviation for Radio-Television Manufacturers Association, now called Electronic Industries Association (EIA).

rumble—1. Turntable rumble. A low-frequency vibration, caused by out-of-round drive wheels, bearings, etc. The vibration is transferred to the pickup head and appears as an undesirable rumble in the sound output. 2. Low-frequency noise from similar causes in a tape transport mechanism.

S

s—Abbreviation for second.

sabin—A unit of sound absorption, used in acoustics. It represents the absorption obtained from 1 square foot of a perfectly absorptive surface.

sapphire—A pure form of corundum (Al_2O_3) used because of its hardness and wearing qualities for tips of phonograph styli and for bearings in watches and other precision instruments.

saturation curve—One type of magnetization curve for a ferromagnetic material.

saturation magnetization—The maximum magnetic strength that a material can attain, regardless of the strength of the magnetizing force applied to it.

saturation noise—Noise obtained on playback of a uniformly saturated tape. Compared with bulk-erased tape, its noise level is about 15 dB higher.

sawtooth current—Current having a sawtooth waveform.

sawtooth voltage—A voltage having a sawtooth waveform.

sawtooth wave—A periodic waveform, shaped like saw teeth. One side of each tooth is usually longer than the other side.

scale—In music, a succession of tones, each bearing a definite ratio to other tones of the scale.

scratch filter—An audio filter that passes low frequencies and suppresses highs, tending to reduce needle-scratch noise.

secondary—The winding of a transformer that provides the output, as contrasted with the primary, to which input voltages or signals are applied.

secondary cell—A voltaic cell in which chemical action is reversible—that is, after being discharged, the cell may have its potential restored by passing current through the cell in a direction opposite to the discharge current.

secondary voltage—The voltage developed across the secondary winding of a transformer.

secondary winding—In a transformer, the winding that delivers output voltage.

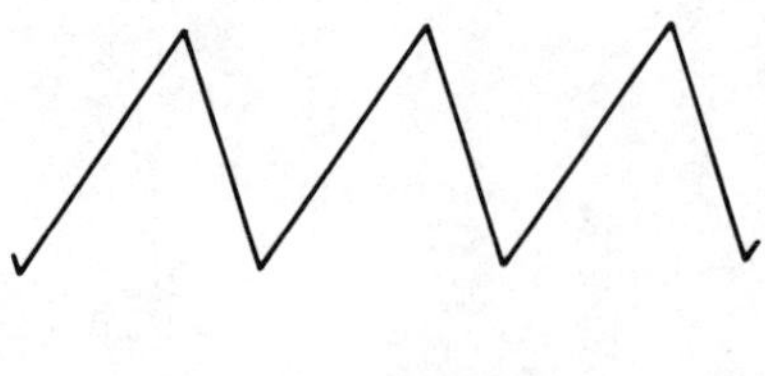

Sawtooth wave.

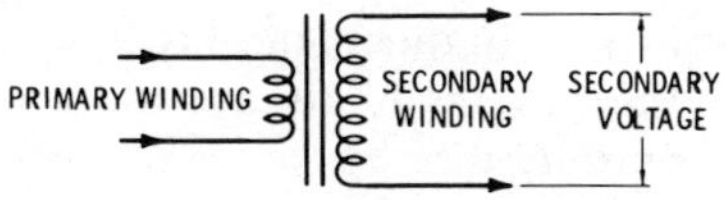

Secondary voltage.

selectivity—The ability of a tuner to accept the signal from one station while rejecting the signal of another only slightly different in frequency.

self-bias—In a vacuum-tube circuit, bias obtained by using the voltage drop caused by current through a resistor in either the grid or cathode lead.

self-demagnetization—A characteristic of permanent magnets whereby the material tends to demagnetize itself because of its own internal fields. This phenomenon is more evident for recorded signals of high frequency or of steep waveforms.

self-generating transducer— A transducer that does not require an external source of electrical excitation for its operation.

self-inductance—That property of a wire conductor that opposes any change in current.

self-threading reel—A tape reel that is easily threaded, without recourse to external aids or threading devices.

semiconductor— An electronic conductor with a resistivity somewhere between that of metal conductors and insulators. An essential difference between semiconductors and metal conductors is that thermal effects at room temperatures generate relatively few free electrons in semiconductors but generate a great many in metals.

semiconductor diode—A device made of n-type and p-type semiconductor material so joined as to give a pn junction. This junction passes current in the forward direction and blocks it in the reverse direction.

semitone—The interval between two adjacent tones in the chromatic musical scale. In the even-tempered system, two successive tones have a frequency ratio equal to the twelfth root of 2. Also called half step.

sensitivity—The ability of a tuner to receive a weak signal and amplify it to usable levels.

separation—In a stereo system, the degree to which the right-channel and left-channel signals can be kept apart.

separation filter—A filter network designed to separate one band of frequencies from another, without losing either band.

separation loss—Loss in output of a tape recorder when the tape does not make intimate contact with the heads.

series circuit—Said of components that are connected end-to-end so that a single path is provided for current.

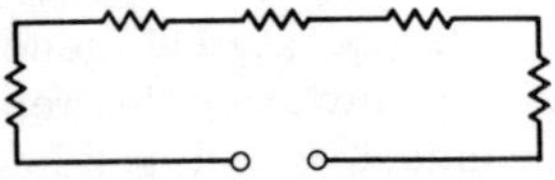

Series circuit.

series motor—A motor having field and armature windings in series. Often called a "universal" motor, it is usually a fractional-horsepower motor.

series-parallel network—A network consisting of series and parallel connections; the variety of combinations is very great.

series resonance—In a circuit having a source of electromotive force in series with inductance and capacitance, series resonance is the condition in which capacitive reactance and inductive reactance are equal and maximum current flows.

series-wound motor—*See* Series Motor.

shaded-pole motor—A single-phase induction motor started by one or more short-circuited stator windings. The short-circuited winding creates a small field displaced in phase from the main winding and this provides the starting torque. This motor type is often used in inexpensive record-player mechanisms.

shading coil—*See* Shading Ring.

shading ring—Heavy copper ring placed around part of the pole piece of small ac motors to produce a starting field.

shaving—In mechanical recording, the process of resurfacing a recording disc by shaving or trimming the surface irregularities.

Shibata stylus—A phonograph stylus of special tip shape developed for playing quadraphonic phonograph discs.

shield—1. A conducting, metallic screen placed around a circuit component and grounded to isolate the component from stray electric fields. 2. A screen of magnetic material placed around a component to divert magnetic fields around the component.

shielded cable—A cable containing one or more conductors surrounded by a separate conductor designed to protect the interior conductors from the effects of undesirable electric fields.

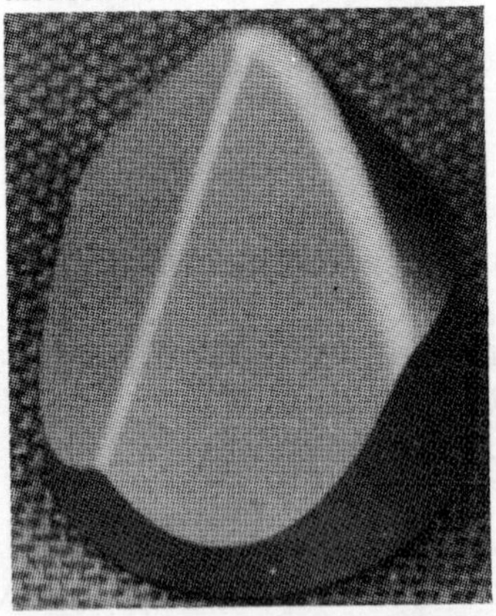

Shibata stylus tip. (Courtesy JVC America, Inc.)

shielded wire—An insulated wire surrounded by a metal shield, usually of fine braided wire.

shock mount—An elastic support used to isolate a system from external vibrations. Often used between a turntable and its cabinet so that room vibrations will not be transmitted to the pickup.

short circuit—Also called a short. A low-resistance conducting path, usually accidental but sometimes intentional, between two points in a circuit.

short-circuit current—The current that a power supply is able to deliver through a very low resistance, such as a short circuit.

shorting switch—A switch in which contact for a new position is made before contact for the previous position is broken. Also called "make - before - break" switch.

shunt—1. A conductor or component placed in parallel with a device or some other part. 2. A precision resistor placed across a meter movement to increase its range.

shunt-wound motor—A direct-current motor having the field circuit and armature circuit connected in parallel.

side thrust—In disc recording and playback, the radial component of force caused by stylus drag and tending to move the pickup arm toward the center of the record.

signal level—A measure of the strength of a signal by comparing its amplitude with the amplitude of an arbitrary reference signal. The ratio of signal level to reference level is often expressed in decibels.

signal-to-noise ratio—In a sound system, or in any component of a sound system, the ratio of the desired output signal to the undesired noise created by the system itself. Usually stated in dB. Abbreviated s/n ratio or snr.

silicon steel—A steel often used in transformer cores and similar devices. It contains 3% to 5% silicon.

silicon transistor—A transistor using silicon as the semiconductor material.

silver-mica capacitor—A capacitor that uses mica for the dielectric and a deposit of silver on the mica for the conducting element.

simple harmonic current—A symmetrical alternating current such as that produced by a commercial ac power system, having a value at any instant equal to its maximum, or peak, value times the sine of the angular rotation of the generator at that instant.

simple sound source—A sound source that radiates sound uniformly in all directions in a free field.

simple steady-state vibration—A periodic motion made up of a sin-

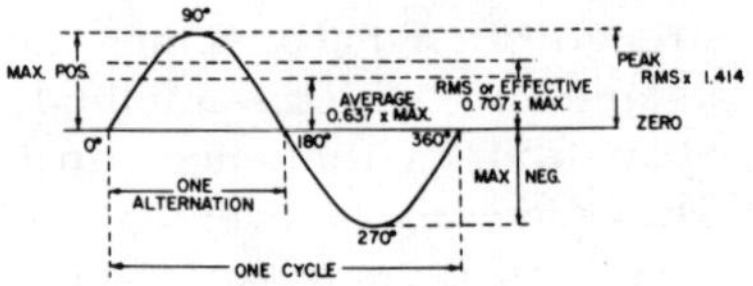

Sine wave.

gle sinusoid without harmonics or other frequencies.

simple tone—Also called pure tone. A musical tone characterized by absence of any harmonics or overtones. Representative of a single sinusoidal frequency.

sine—In a right triangle, the sine of an angle is equal to the side opposite the angle, divided by the hypotenuse.

sine wave—The waveform obtained when the graph of y= sine x is plotted, where y is measured along the vertical axis and x is measured along the horizontal axis in radians or degrees. In electronic applications, y usually represents voltage or current.

single-button carbon microphone—A microphone having a small button-like container filled with carbon granules on one side of a diaphragm. As the diaphragm is vibrated by sound waves, the carbon granules are alternately compressed and expanded, causing the microphone resistance to change in step with the signal.

single-ended amplifier—An amplifier having only one tube or transistor at the output stage. Hence, the output is unbalanced to ground. If more than one tube or transistor is used at the output stage they are connected in parallel, so that output is still unbalanced with respect to ground.

single-groove stereo—*See* Monogroove Stereo.

single-track recorder—A recorder that records or plays only one track on the tape.

sinusoid—A sine curve.

sinusoidal—Having the shape or values of a sine curve.

slip—In a synchronous motor, the difference between actual operating speed of the motor and the theoretical speed if the motor were in exact step with the applied voltage.

small-signal characteristics—The characteristics of an amplifier operating under conditions that result in linear amplification.

smoothing choke—A choke used to smooth out variations in the output of a power supply.

SMPTE—Society of Motion Picture and Television Engineers.

s/n or snr—Abbreviation for signal-to-noise ratio.

soft magnetic material—Material that, once magnetized, is easily demagnetized.

solenoid—A single layer of wire wound in the form of the surface of a cylinder, but not necessarily circular.

solid state—Circuits, amplifiers, etc., that use semiconductor components such as transistors and integrated circuits instead of vacuum tubes.

solo manual—*See* Swell Manual.

sone—A subjective unit of loudness. Based on the listener's own threshold of hearing, it is the loudness of a 1000-hertz tone that is 40 dB above that threshold.

sonic frequencies—Same as audio frequencies: sound vibrations ranging from about 15 hertz to 20,000 hertz.

sound—1. The sensation experienced through the sense of hearing. 2. Mechanical radiant energy transmitted by vibrations in an elastic medium such as air or other materials. 3. Recorded auditory material.

sound absorption—The interception and dissipation of sound waves, usually with the generation of heat by the function of moving molecules.

sound-absorption coefficient—A figure expressing the relative effectiveness of a material in absorbing sound waves.

sound analyzer—A device for measuring the amplitude and frequency of the components of a complex sound.

sound intensity—A measure of the strength of a sound wave by its rate of transmitting sound energy through a unit area. The common unit is the erg per second per square centimeter. Another unit is the watt per square centimeter.

sound level—The overall loudness of sounds, based on the judgment of equal loudness of pure tones. The level is expressed in dB with respect to 0.0002 microbar (accepted threshold of hearing).

sound-level meter—An instrument for the measurement of noise and sound levels. The instrument usually includes a microphone, an amplifier, an output meter, and frequency-weighting networks.

sound-on-sound—A method of recording new program material on the same track as previously recorded material, without erasing the first program. Thus the two recordings are heard simultaneously on playback.

sound-power level—The ratio in dB of the sound power of a source to a standard reference level of 10^{-13} watt.

sound-pressure level—The level in decibels of a sound pressure aas compared to a standard reference level. Two reference levels are in common use: 2×10^{-4} microbar and 1 microbar.

sound velocity—The speed with which sound vibrations are transmitted through a medium such as air, water, steel, wood, etc. The velocity is affected slightly

Materal	*Feet per second*
Aluminum	16,740
Brick	12,500
Copper	11,670
Carbon dioxide	846
Glass	16,400
Iron	16,830
Lead	4026
Pine (direction of grain)	10,900
Silver	8553
Water	4728
Steel	16,500

Sound velocity in various mediums.

by the temperature of the medium or by its pressure (if medium is a gas). Usually expressed in units of feet per second.

south pole—The pole of a magnet where the magnetic lines of force are assumed to enter.

speaker—Abbreviated spkr. Also called loudspeaker. A transducer that converts electrical impulses into motion of a diaphragm or similar sound-radiating element. Air waves set in motion by this element create the sensation of sound in the ears of the listener.

speaker efficiency—Ratio of acoustical power radiated by a speaker to the electrical power supplied to the voice coil.

speaker impedance—The vector combination of resistance and reactance of the voice coil of the speaker. This varies with frequency, so it is usually stated for 400 Hz.

speaker system—A combination of speakers, baffles, horns, and dividing networks, used to couple an audio amplifier to the listening area.

speaker voice coil—A coil of conductor material, wound on a lightweight form fastened to the speaker cone. Audio signal currents through the coil create a magnetic field that reacts with the stationary magnetic field of the speaker to produce movement of the cone, thus creating sound waves.

specific acoustic impedance—A measure of the opposition to sound transfer offered by a unit area of a medium. It consists of a real and an imaginary component and is the ratio of sound pressure to particle velocity at a point in the medium.

specific acoustic reactance—The imaginary component of the specific acoustic impedance.

specific acoustic resistance—The real component of the specific acoustic impedance.

specific resistance—*See* Resistivity.

speech amplifier—A voltage amplifier designed especially for use with a microphone.

speech clipper—A circuit for limiting the amplitude of a speech signal to allow an increase in modulation percentage of an amplitude-modulated transmitter.

speech level—The energy contained in speech as measured by a level indicator such as a VU meter.

speed of sound—The rate of travel of sound waves through a particular medium. The speed varies, depending on the medium, temperature, altitude, air pressure, etc. For air at normal sea-level conditions, the speed of sound is about 1130 feet per second. In many solids and liquids, the speed is much greater than in air, ranging from 4794 feet per second in water to 18,050 feet per second in glass.

spider—A flexible, washer-shaped component used to center the voice coil of a speaker on the magnetic pole piece. It is designed to permit free travel of the voice coil in an in-and-out direction while limiting travel in a sidewise direction.

spindles—In tape recording, the posts on which the tape reels are centered on the recorder; in phonographs, the upward projecting shaft in the center of the turntable.

spkr—Abbreviation for speaker.

spl—Abbreviation for sound-pressure level.

splicer—A device for making splices.

splice, tape—A joint between two pieces of tape.

splicing tape—A pressure-sensitive, nonmagnetic tape for splicing magnetic tape. It has a special adhesive coating that will not soften or "bleed."

spread groove—In a record consisting of a number of separate selections, the unmodulated groove that leads quickly from one selection to the next.

SQ—Matrixing system developed by CBS Laboratories for quadraphonic disc recording and playback.

square wave—A repeating, rectangular-shaped wave with positive and negative excursions equal in amplitude and duration.

squawker—Popular slang term for the midrange speaker of a three-way speaker system.

squeal, tape—Audible tape vibrations caused by pulling tape past dirty or glazed pressure pads.

stabilization network—In an amplifier with negative feedback, a network used to prevent oscillation.

stacked heads—In stereophonic tape recorders, an arrangement that places the right and left channel heads one over the other, or "stacked."

staggered heads—As opposed to "stacked heads," an alternate arrangement of the recording heads in a stereophonic tape recorder. One head is displaced from the other by 1 $^{7}/_{32}$ inches.

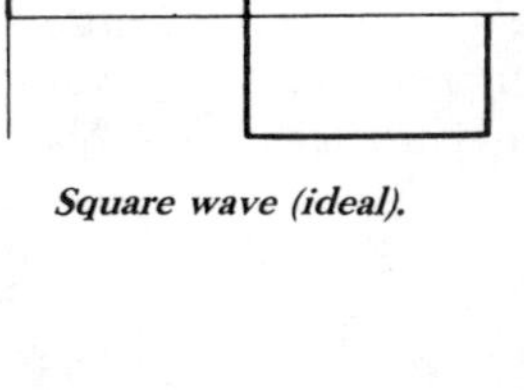

Square wave (ideal).

standard microphone—A microphone having a response that is known under the conditions for which it is to be used. Thus, it can be used as a standard of comparison when testing microphones of unknown performance.

standard noise temperature—A standard reference temperature for noise measurements. The accepted standard is 290° K.

standard pitch—The standard pitch for the musical scale, which places the note A at 440 hertz.

standard test-tone power—A power of 1 milliwatt (0 dB) at 1000 hertz.

steady state—A condition in which values sometimes subject to variation (voltage, current, and resistance) remain essentially constant, after initial transients or fluctuations have subsided.

steady-state oscillation—Oscillation that remains substantially constant in frequency and amplitude.

step-down transformer—A transformer having more turns in the primary than in the secondary—hence, the output voltage is less than the input voltage.

step-up transformer—A transformer having more turns in the secondary than in the primary—hence, the output voltage is greater than the input voltage.

stereo—A prefix meaning three-dimensional. Also a common abbreviation for stereophonic.

stereo adaptor—A device used to convert two identical sets of monophonic equipment to operation as a single stereophonic system.

stereo amplifier—An audio-frequency amplifier for a stereophonic system, having two or more channels.

stereo cartridge—A phonograph pickup for playback of stereophonic recordings. The stylus mechanism has high compliance in horizontal and vertical directions and produces output for motion components in both directions.

stereocasting—Broadcasting over two sound channels to provide stereophonic reproduction. Methods for achieving this result are simulcasting, multicasting, and multiplexing.

stereo head—Tape-recording head used for recording stereo pro-

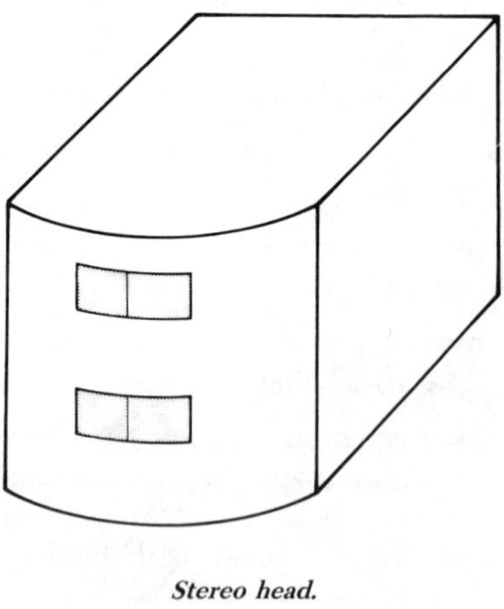

Stereo head.

grams on tape. May have either two or four record gaps.

stereo microphones—Two or more microphones so spaced as to provide three-dimensional pickup for stereophonic recording.

stereophonic — Multichannel, three-dimensional sound, designed to produce a sense of separation like that of the original sound sources.

stereophonic separation—The degree to which the signal in one channel of a stereo amplifier system is free from signals relegated only to the other channel.

stereophonic sound—Dimensional or directional sound reproduction achieved through the use of two or more sound channels heard simultaneously through two or more speakers.

stereophonic system—A complete sound-reproducing system, both recording and playback, that includes microphones, amplifier channels, and speakers (or headphones), with the result that original three-dimensional sound is reproduced.

stereo pickup—*See* Stereo Cartridge.

strain—Physical deformation or mechanical displacement of an object resulting from applied stress (mechanical force). In electronics, strain finds application in certain types of phonograph pickups.

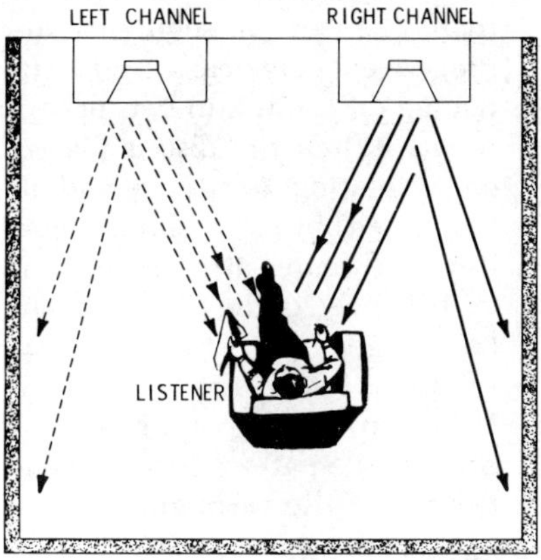

Stereophonic system.

strain gauge—A transducer made of material that changes in resistance when subjected to strain. This property makes it useful for converting force, pressure, etc., into an electrical signal.

strain pickup—A phonograph pickup that uses the principle of the strain gauge to convert record-groove modulation into an electrical output.

stroboscopic disc—A disc having alternate black and white radial segments arranged in circular fashion around the center of the

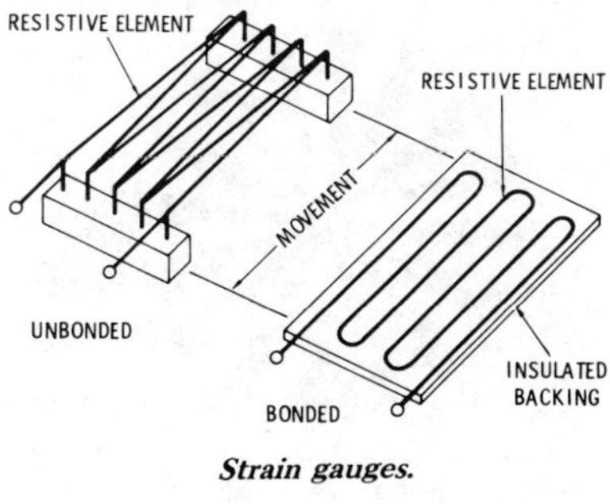

Strain gauges.

disc. The disc may have more than one set of segments arranged concentrically and containing different numbers of segments. When the disc is placed on a rotating turntable and illuminated by a flashing light, usually flashing at line frequency, the speed of the turntable can be judged by the apparent motion of the strobe pattern. Many turntables have a permanent strobe pattern located near the rim of the turntable.

stylus—The tiny, needle-like part of a phonograph pickup or record-cutting head that follows or cuts the groove of the record. The tip of the stylus is made of various hard materials such as diamond, sapphire, or Stellite.

stylus drag—The friction between stylus and record groove.

stylus force—The downward force exerted on the record by the stylus. Also called stylus pressure, needle pressure, and tracking force.

stylus pressure—*See* Stylus Force.

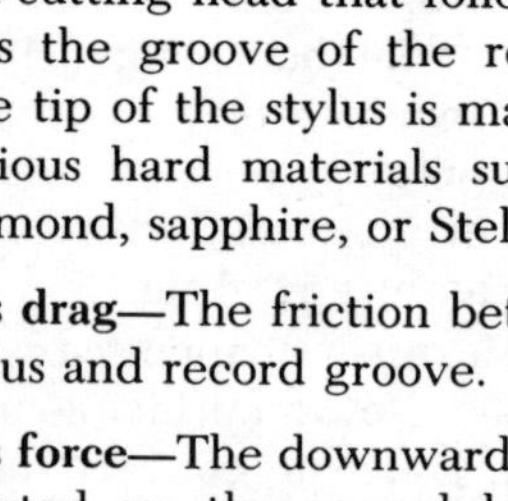

Stroboscopic disc.

subharmonic—A frequency component that is an integral submultiple of a fundamental tone—that is, one-half, one-third, one-fourth, etc., of the fundamental frequency.

submultiple resonance—Resonance at a subharmonic of the fundamental exciting frequency.

subsonic frequency— Obsolete term for infrasonic frequency, which see.

subsonic speed—Speed less than the speed of sound.

sum channel—A channel that accepts material from right and left channels of a stereo system and combines it in a form that passes as an accurate monophonic representation of the program and that can be recorded or transmitted as such.

summation tone—Another name for combination tone. When two tones are combined in a nonlinear device, a third tone, the summation tone, equal in frequency to the sum of their frequencies, is produced.

supersonic—Faster than sound (approximately 1130 feet per second, or 750 miles per hour). Speeds equal to or faster than the speed of sound are often given Mach numbers, which show the ratio of their speed to the speed of sound in air. Thus, Mach 1 is equal to the speed of sound,

Mach 2 is twice the speed of sound, and so on.

supersonic frequency—*See* Ultrasonic Frequency.

supply reel—On a tape recorder, the reel from which tape is removed during recording or playback operations.

surface noise—In playback of disc recordings, noise caused by minute irregularities in the surface of the record material, dirt, scratches, etc. Also called needle scratch.

surround—The outer edge of a speaker cone that fastens to the speaker frame, or basket.

susceptance—The imaginary component of admittance. It is the reciprocal of reactance.

susceptibility—The ratio of induced magnetization to the inducing magnetic force.

sustain—In an electronic organ, a circuit design feature that prolongs the sound of a note for a predetermined period after the key has been released.

sweep-frequency generator—A signal generator whose output sweeps through a continuous range of frequencies, repeating the sweep a number of times per second. Often used in conjunction with an oscilloscope to display response curves.

sweep-frequency record—For audio frequencies, the mechanical equivalent of the sweep-frequency generator, with the sweep-repetition rate typically 20 sweeps per second. The sweep signal is recorded on a disc that can be played on a record player. Used for testing pickups, amplifiers, and speaker systems.

swell manual—The upper manual of a two-manual organ. Normally used to play the melody. Also called solo manual.

synchronous motor—An induction motor whose rotor is in step (synchronized) with the alternations of the ac voltage applied to it.

synchronous speed—The speed at which a synchronous motor runs. To find the speed of a synchronous motor in revolutions per minute, divide the frequency of the driving voltage in hertz by the number of poles, then multiply by 120.

synthesizing—A process for obtaining four-channel effects from existing two-channel stereo material. In subjective tests, listeners have agreed that a two-channel stereo signal is enhanced when fed to a matrix decoder.

T

T—Abbreviation for the prefix "tera."

take-up reel—The reel on which tape accumulates during record or playback of a tape recorder.

tape cartridge—An enclosed container for magnetic tape that can be inserted directly into a tape player without the necessity of manual threading or handling.

tape deck—The basic assembly of a tape recorder, including the mechanism for moving the tape and a head assembly. Some tape decks include a preamplifier or preamplifiers (stereo).

tape drive—The mechanism for moving the tape past the heads of a tape recorder.

tape guides—Grooved pins or posts mounted on either side of the head assembly. They position or guide the tape so that it passes the heads with proper alignment.

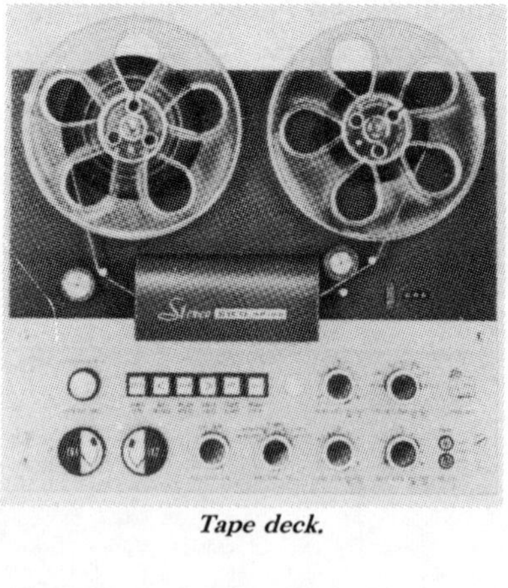

Tape deck.

tape lifters—A mechanism for lifting the tape away from the recorder heads as the recorder is in "Fast Forward" or "Rewind" position. This protects the heads from unnecessary wear.

tape loop—A length of magnetic tape with ends joined to form an endless loop. When played on the proper equipment, it provides for continuous repetition of a message.

tape magazine—*See* Tape Cartridge.

tape, magnetic—A thin plastic ribbon coated wth iron oxide, which will accept and retain magnetization placed on it by the recording head.

tape monitor—Provision for checking quality of a tape recording as it is being made, by directly listening to the recorded signal with an extra playback head.

tape player—Unit for playback of tapes, but with no provision for recording.

tape, prerecorded—Tape that has program material already recorded on it. A more appropriate term would be "recorded tape."

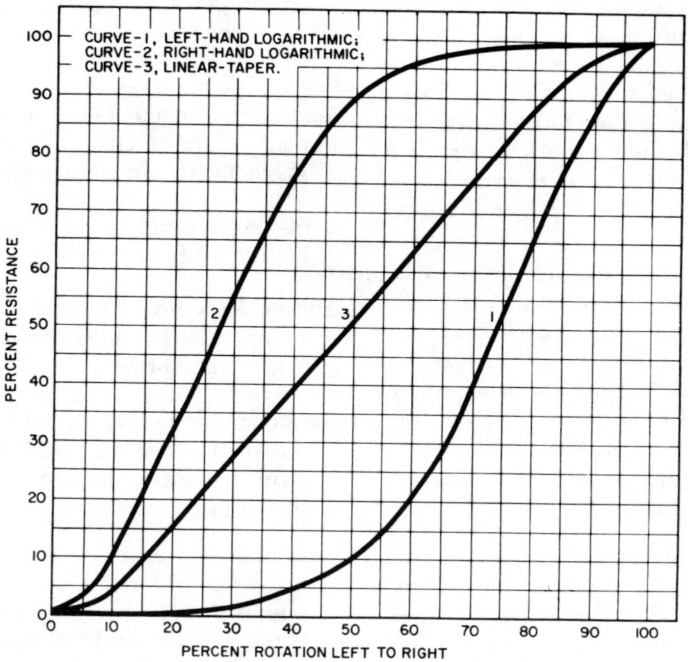

Taper curves for potentiometer.

taper curves—Curves showing the resistance characteristic of various potentiometers such as are used in audio circuits as tone controls, volume controls, equalizers, etc.

tape recorder—In its most complete form, a machine for recording and playback of voice, music, and other audio-frequency material. Sound is picked up by microphone, converted to an elec-

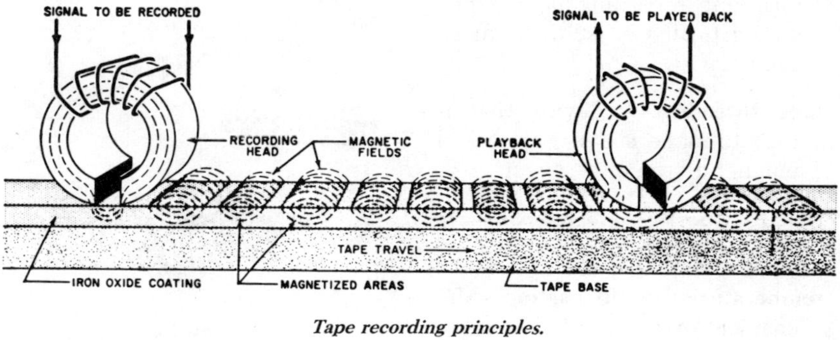

Tape recording principles.

trical signal, amplified, and converted to a varying magnetic field, which is permanently impressed as a magnetic pattern on the recording tape. During playback, the magnetic pattern is reconverted to an electrical signal, amplified and applied to a speaker system.

tape skew—Failure of the tape to follow a straight-line path across the record-playback heads. This departure from normal can cause a time displacement between signals on different tracks and amplitude differences between outputs from individual tracks.

tape speed—The rate at which tape travels past the heads during record or playback. Speeds are usually multiples of 7½ inches per second—that is, 15, 7½, 3¾, 1⅞. Tape speeds in common use for home-type recorders are 7½, 3 ¾, and 1⅞ inches per second.

tape-speed variations—*See* Flutter.

tape splicer—A device for splicing magnetic recording tape either automatically or semiautomatically.

tape threader—A device that is used to make it easier to thread magnetic tape onto a reel.

tape-transport mechanism—*See* Motor Board.

temperature compensation—Mechanical or electronic means of offsetting the effects of temperature changes on a particular device.

temporary magnet—A magnet made of material that is easily magnetized and demagnetized.

tera—Prefix indicating a multiplier of 10^{12}. Abbreviated T.

terminal—A circuit wiring component used in connecting one conductor to another.

terminal board—Also called terminal strip. An insulating base or strip containing terminals used for wiring purposes.

terminal lug—A fitting that is attached or soldered to the ends of conductors so that they can be easily connected to a terminal board or other points in a circuit.

terminal strip—*See* Terminal Board.

test clip—A spring clip of conductive material fastened to wires or test leads for quick temporary connections to electrical circuits.

Terminal boards.

test jack—A jack connected to a circuit so that test equipment can be quickly attached.

test lead—A flexible insulated lead used mainly for connecting test equipment to a circuit.

test prod—A sharp metal tip, usually on the end of an insulated handle, used for making quick temporary connections to a circuit.

test record—A phonograph disc containing recorded signals that are suitable for testing amplifiers, pickups, speaker systems, etc.

thermal microphone—A microphone having its action based on the heating effect of sound waves. The heating effect varies the resistance of an electrically heated conductor, and the variation of resistance in turn produces a variation in current output.

thermal telephone receiver—Also called thermophone. A transducer that varies in temperature according to variations in electrical input. The temperature variations in turn create sound waves by the resulting expansion and contraction of the surrounding air.

thermistor—A semiconductor device having an electrical resistance that varies with temperature. It has a high negative temperature coefficient.

thickness vibration—In a piezoelectric crystal, vibration in the direction of its thickness.

third harmonic—A component of a complex waveform that is three times the fundamental frequency of the waveform.

third-harmonic distortion—The third-harmonic signal generated in an amplifier when a sine-wave signal of fundamental frequency is applied to the input. To express the distortion in percentage, the rms value of third harmonic is divided by the rms value of the fundamental output and the quotient is multiplied by 100.

threading slot—A slot in the cover plate of a recording-head assembly that guides the tape into position during the tape-threading operation.

three-channel stereo—A stereo recording process or system that uses three separated microphones for recording and three sound channels for playback.

three-way speaker system—A sound reproducing system that uses three separate speaker units, each designed to reproduce a specific band of the audio frequencies (high, middle, and low frequencies). These units are known as the tweeter, squawker, and woofer, respectively.

threshold of audibility—Also called threshold of hearing. As a sound is increased from a level below audibility, the level at which it first becomes audible. Usually expressed in decibels with reference to 0.0002 microbar.

threshold of feeling—Sometimes called the threshold of pain. The sound level at which hearing is accompanied by a feeling of discomfort. Usually expressed in decibels relative to 0.0002 microbar.

threshold of hearing—*See* Threshhold of Audibility.

throat—The small cross-sectional area of a horn near the driver unit.

throat microphone—A microphone worn in contact with the throat so that vibrations of the larynx of the user are picked up. It is used in locations of high background noise to eliminate pickup of these noises.

thump—Descriptive term for low-frequency transient disturbances in a sound system or component.

THz—Abbreviation for terahertz (10^{12} hertz).

timbre—The musical quality of a tone that enables a listener to distinguish one musical instrument from another—e.g., a violin from a flute, etc. Timbre derives mainly from the frequency and intensity of the harmonics present in a tone.

time constant—When referred to a resistance-capacitance circuit or a resistance-inductance circuit, the time constant is the time when the change in voltage or current is 63.2% complete. The time in seconds is equal to R times C in the one case and L divided by R in the other case. R is resistance in ohms, C is capacitance in farads, and L is inductance in henrys.

time-division—The sharing of a single channel by two or more information signals by allocating successive time intervals to each signal.

tone—1. A sound wave capable of producing the auditory sensation of pitch. 2. The sound sensation of pitch so produced.

tone arm—The arm of a record player that extends over the record and holds the pickup cartridge. Most tone arms are pivoted but a few have been designed to hold a fixed position while a small carriage containing the pickup travels the length of the arm from margin to center of the record.

tone burst—A short wave train, 50 to 5000 microseconds in length, of a single sine-wave frequency, useful for testing the transient response of speakers.

tone control—A control used for changing the frequency response of an audio system to suit the listener's personal tastes. In its most common form, it is part of a resistance-capacitance network designed to either increase

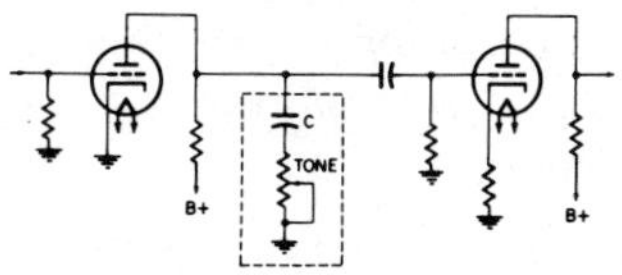

Tone control.

or decrease bass or treble response.

tone generator—1. A generator for producing audio-frequency signals suitable for testing audio-frequency equipment. 2. A circuit used in electronic organs for generating basic tones that are further modified by waveshaping or frequency-multiplying circuits to get the various tones produced by the organ.

total distortion—The sum of all forms of distortion generated in a device.

total harmonic distortion—In an amplifier system, the ratio of the total rms harmonic power output to the total rms power output of the fundamental frequency when a single sine-wave signal of the fundamental frequency is applied to the input.

tracing distortion—In a disc recording system, distortion arising from failure of the reproducing stylus to follow the exact path of the modulated groove.

track—In tape recording, the magnetized path laid down by the recording head on a recording tape.

track, full—A track that covers nearly the full width of the recording tape.

tracking—In disc recording, the exactness with which the reproducing stylus follows the modulated groove of the record.

tracking error—In lateral disc recording, the angle between the vibration axis of the reproducing stylus mechanism and a plane tangent to the unmodulated record groove at the point of stylus contact and perpendicular to the surface of the record.

tracking force—*See* Stylus Force.

track, quarter—A track that occupies slightly less than one-quarter of the tape width; i.e., four tracks on one tape.

track width—The width of the track that a recording head records on the magnetic tape. Common track widths are 48 and 50 mils.

trailing edge—The edge of a step pulse that occurs last in time, as opposed to the leading edge, which occurs first.

transcription—Also called electrical transcription. A disc recording of program material used in radio broadcasting, formerly recorded on 16-inch discs, now recorded on 10- or 12-inch discs, using microgroove techniques.

transducer—A device that converts energy from one form to another (e.g., a phonograph pickup converts mechanical energy to electrical energy).

transformer—An electrical device that transfers electrical energy from one circuit to another by means of electromagnetic induction between two or more windings in a common magnetic field.

transient—A short term impulse or disturbance in a system, characterized by a steep wavefront and fast decay.

transient distortion—Distortion caused by inability of a system to reproduce transient inputs faithfully.

transient oscillation—Oscillation of short duration caused by switching transients, signals with steep wavefronts, etc.

transient response—The response of an amplifying system to a step function input such as a square wave or a unit pulse. Response is evaluated in terms of rise time, fall time, overshoot, undershoot, preshoot, and ringing.

transistor—An active semiconductor device, usually made of silicon or germanium and having three or more electrodes. The three electrodes in a simple transistor are emitter, base, and collector.

transition frequency—In a disc-recording system, the frequency at which the recording characteristic changes from constant amplitude to constant velocity. Also called crossover frequency.

translation loss—Also called playback loss. In mechanical recording, signal loss occurring when excursions of the reproducing stylus differ in amplitude from recorded amplitudes on the record.

transverse magnetization—Magnetization in a recording tape perpendicular to the normal direction of travel.

treble—The higher frequencies in harmonic music or voice; generally considered the frequencies above 261.63 hertz (middle C).

treble boost—An increase in the amplitude of higher audio frequencies in comparison to the amplitude of the lower frequencies.

tremolo—Low-frequency variation or modulation of the amplitude of an audio tone.

triaxial cable—A type of coaxial cable containing three conductors.

triaxial speaker—A speaker assembly consisting of three independently driven units mounted coaxially.

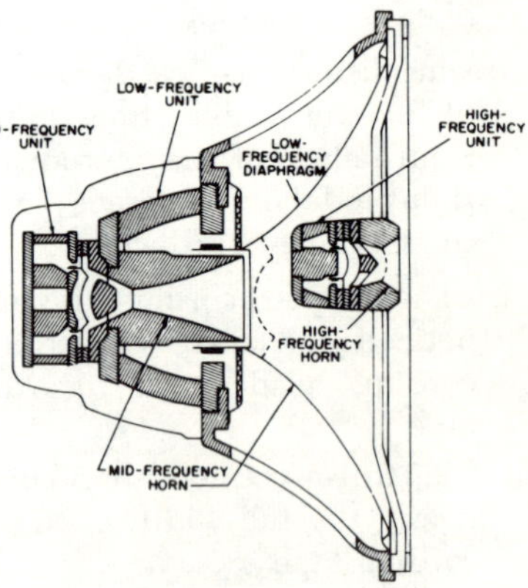

Triaxial speaker.

true power—Actual power consumed by a component in a circuit. It is equal to I^2R, where I is the current through the component and R is the resistive part of the impedance of the component.

tube noise—Noise in an amplifier circuit contributed by a vacuum tube (all noises not directly attributed to applied signal, such as shot noise, thermal agitation, etc.).

tuned—1. In electronics, adjusted to be resonant at a particular frequency. 2. In musical instruments, having the tone-generating components adjusted to produce their designated frequencies.

tuned amplifier—An amplifier having a tuned circuit for a load.

tuned circuit—A circuit containing both inductance and capacitance and having provision for adjustment of one or the other or both.

tuner—1. The section of a radio or tv receiver designed for receiving and tuning rf signals and delivering rf or i-f signals to the next section of the receiver. 2. A separate unit designed for performing the first functions of a receiver and delivering rf, i-f, or demodulated signals to succeeding amplifier and power output units.

tuning eye—Common name for a cathode-ray tuning indicator.

tuning fork—A two-pronged fork made of some highly elastic metal and used to generate a pure audio tone. Can be activated either by striking a prong or by an electronic drive circuit magnetically connected to the prongs.

tuning indicator—A device that indicates when a tuned circuit is accurately tuned to an incoming signal. Normally either a meter or some form of cathode-ray indicator.

tuning meter—A direct-current meter used to indicate when a tuned circuit is properly adjusted to the frequency of an incoming signal. It does this by indicating either a minimum or a maximum, depending on the design of the tuned circuit.

turnover cartridge—A phonograph cartridge designed to play both large- and fine-groove records by means of two sizes of stylus tips mounted either on opposite sides of the cartridge or on opposite sides of a turnover stylus.

turnover frequency—In disc recording, the dividing-line frequency between low frequencies

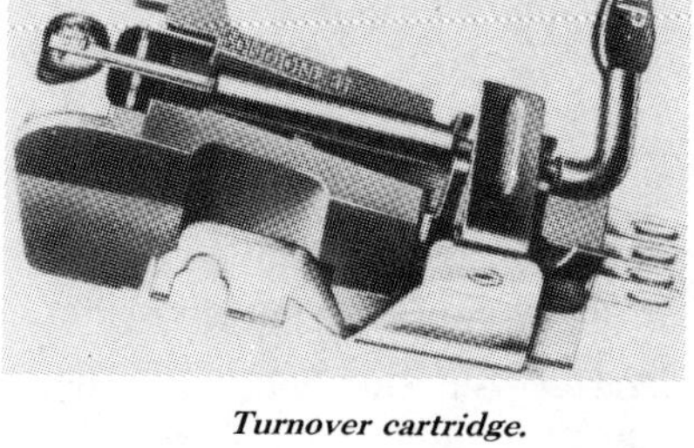

Turnover cartridge.

recorded at constant amplitude and high frequencies recorded at constant velocity.

turns ratio—In a transformer, the ratio between the number of turns in the primary winding and the number of turns in the secondary winding.

turntable—The revolving platform used for recording or playing phonograph records.

turntable rumble—*See* Rumble.

tweeter—Common name for a high-frequency speaker unit. A speaker used to reproduce the high-frequency audio band, particularly, those above 3000 hertz.

two-track recording—In tape recording, a recording of two channels of sound, either as a stereo pair in one direction or as two separate monophonic tracks (usually in opposite directions).

two-way system—A speaker system consisting of a unit for reproducing all frequencies above a certain frequency and another unit for frequencies below that frequency. A divider network of some sort is used to separate the highs and the lows before they are applied to each speaker.

U

uhf—Abbreviation for ultrahigh frequency.

ultralinear amplifier—Name applied to a certain high-fidelity amplifier circuit requiring a specially wound output transformer.

ultrasonic—Frequencies above those of audible sound.

ultrasonic generator—A device for generating vibrations at frequencies above the range of human hearing.

ultrasonics—The study and application of sound waves having frequencies above the limit of audibility (15 kilohertz).

ultrasonic transducer—A device used for converting electrical signals of ultrasonic frequency to mechanical vibrations of the same frequency. Materials suitable for this purpose are some of the piezoelectric materials (e.g., quartz and barium titanate) and some magnetostrictive materials (e.g., nickel).

ultrasonic waves—Waves with a frequency in the ultrasonic range.

undamped natural frequency—The frequency at which a body will vibrate if subjected to a momentary displacement impulse, in the absence of any damping resistance.

undamped oscillation—Oscillation that maintains a constant amplitude during its lifetime.

undamped wave—A waveform of constant amplitude.

undercutting—In disc recording, the cutting of a groove that is too shallow or that has insufficient lateral excursion to reproduce properly.

underdamped—The condition of not having enough damping to prevent oscillation on application of a step signal or steep transient.

undistorted wave—A periodic wave whose amplitude at all points bears a direct unchanging ratio to the amplitude of the original wave and whose frequency and phase are unchanged.

unidirectional microphone—A microphone that has an effective response only to sound waves arriving in a single solid angle of one hemisphere or less.

unidirectional transducer — A transducer that responds only to stimuli applied in one direction from a zero or rest position.

unit-area acoustic impedance—*See* Specific Acoustic Impedance.

unit electrical charge—The electrical charge that will exert a force of one dyne on an equal and like charge at a distance of one centimeter in a vacuum.

unit magnetic pole—The strength of a unit magnetic pole is such that two identical poles separated by a distance of 1 centimeter exert a force of 1 dyne between them.

unity coupling—A magnetic coupling coefficient of 1. This indicates that all the flux lines of one winding link the turns of the other winding.

unity power factor—In an ac circuit, a power factor of 1.0. This occurs when the voltage and current are in phase (a purely resistive circuit or a reactive circuit at resonance).

universal motor—A series-wound motor capable of operating either on dc or on single-phase ac. This type is often found in small household appliances such as sewing machines, food mixers, blenders, can openers, etc.

universal output transformer—An output transformer having a number of taps on the secondary winding. In use, the proper tap is chosen to match the output impedance of the output stage to the impedance of the speaker used.

unmodulated—Lacking modulation. In broadcasting, a carrier wave with no applied signal. In disc recording, a blank groove.

unmodulated groove—Also called a blank groove. In mechanical recording, the groove made in the medium with no signal applied to the cutter.

V

V—Symbol for volt or voltmeter.

vacuum-tube voltmeter—A voltmeter using the amplifying and high-impedance properties of vacuum tubes to obtain a sensitive yet rugged meter with low loading characteristics.

variable-area recording—A method of sound recording using a photographic process. The density of exposure is held constant while the area of exposure is varied in agreement with sound variations.

variable-density recording—A method of sound recording using a photographic process. The area of exposure is held constant while the density of exposure is varied in agreement with sound variations.

variable-inductance pickup—A phonograph pickup in which the stylus movement varies the inductance of the pickup and this, in turn, causes the electrical output of the pickup to vary.

variable-reluctance microphone—A microphone in which the vi-

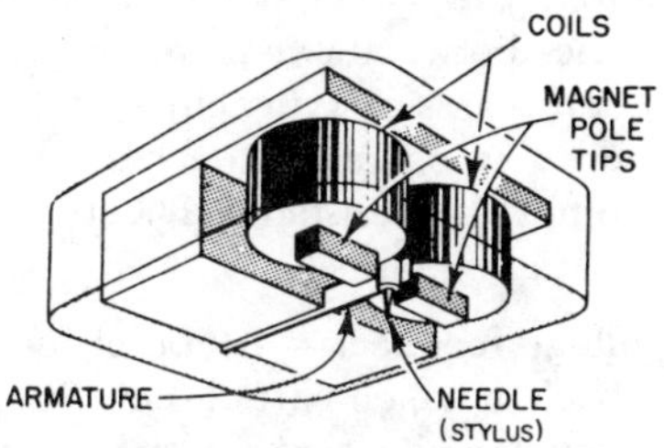

Variable-reluctance pickup.

brating member produces variations in the reluctance of a magnetic circuit.

variable-reluctance pickup—A phonograph pickup that produces a signal by means of variations in the reluctance of a magnetic circuit.

variable-resistance pickup—A phonograph pickup in which stylus vibration produces a variation in the resistance of the pickup.

Variomatrix—A matrix system for four-channel sound developed by Sansui Corporation of Japan.

velocity microphone—A microphone whose output corresponds to the instantaneous velocity of the particles in the impressed sound wave. It has essentially a bidirectional response characteristic.

vertical compliance—The ease with which a reproducing stylus can move in a vertical direction when playing a disc recording. Usually measured in dynes per cm.

vertical groove displacement—In 45/45 stereo disc recording, the stylus displacement when the system is driven by right and left signals of equal amplitude and 180° phase difference.

vertical-lateral recording—A stereo disc recording technique that records one signal laterally, as in monophonic recording, and the other signal vertically, as in hill-and-dale recording.

vertical recording—Also called hill-and-dale recording. A type of disc recording that modulates the groove in a vertical direction rather than in a horizontal direction.

vhf—Abbreviation for very high frequency.

vibration pickup—A microphone designed to respond to mechanical vibrations rather than to sound waves. It is placed in direct contact with the sound source (e.g., the contact microphone used as a guitar pickup).

vibrato—Frequency modulation of a musical tone at a slow rate (usually 5 to 7 Hz).

viscous-damped arm—A phonograph pickup arm whose movement is damped by a highly viscous fluid, usually a silicone fluid. Damping helps to eliminate arm resonances and protects record grooves and stylus in case the arm is dropped.

vlf—Abbreviation for very low frequency.

voice coil—Also called speaker voice coil. A coil of wire or metal ribbon attached to the speaker diaphragm and free to move in the air gap of the speaker pole pieces.

voice filter—A parallel-resonant filter circuit connected in series with a speaker or a number of speakers to improve the speaking qualities of the male voice. Resonant frequency is adjusted to lie between 125 and 300 Hz.

voice frequency—Commonly, the audible range of frequencies (32–15,000 Hz). In telephony, the voice range for maximum intelligibility is the range from 100 to 3500 hertz.

volt—The unit of electromotive force. It is the force required to produce a current of 1 ampere through a resistance of 1 ohm.

voltage—1. Electric potential difference between two points in a circuit. 2. The electromotive force of a source supplying electrical energy to a circuit.

voltage amplification—Ratio of the voltage output of a transducer or amplifier to the voltage input. Also called voltage gain.

voltage amplifier—An amplifier whose main function is to increase the voltage applied to its input.

voltage divider—A tapped resistor or series of resistors connected across a voltage source. Different fractions of the total voltage are available at the various taps.

voltage drop—The potential difference between two points in a circuit that is carrying current. The drop is equal to the current I times the resistance R between the two points.

voltage feedback—A type of feedback in which part of the voltage developed across the load impedance is applied to some preceding point in the circuit.

voltage gain—*See* Voltage Amplification.

voltage ratio (of a transformer)—The ratio of the primary voltage to the secondary voltage. It is equal to the turns ratio.

voltage regulator—A circuit component or a combination of several components that tend to hold the voltage of a source connected to the regulator at a constant value.

voltage-sensitive resistor—A resistor, the value of which varies with the voltage applied to it.

voltmeter—An instrument for measuring the potential difference between two points in an electrical circuit or across the terminals of a voltage source.

voltmeter-ammeter—A combination of voltmeter and ammeter functions in a single case.

volt-ohm-milliammeter — Often called vom. A test instrument used for measuring voltage, resistance, and current.

volume—A term loosely used to indicate the relative intensity or apparent loudness of a sound.

volume compression—Also called automatic volume compression. The limiting of the volume range to about 30 to 40 decibels at the transmitter, to permit a higher average percentage modulation, without overmodulation. Also used in recording to raise the signal-to-noise ratio.

volume control—A control, usually a variable resistance, used for adjusting the loudness of a receiver or an amplifier.

volume expansion—*See* Automatic Volume Expansion.

volume indicator—An indicating device for monitoring the volume of complex audio waves such as that corresponding to speech.

volume indicator meter—Abbreviated VI meter. An instrument for reading audio-frequency power. The term is generally associated with meters calibrated in decibels.

volume limiter—A device in which the gain is automatically reduced whenever the input exceeds a certain value. In this manner, the output is prevented from rising above a certain level.

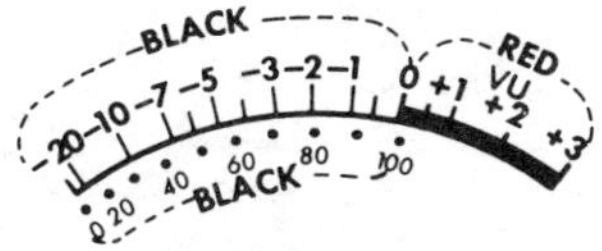

Recording and test equipment.

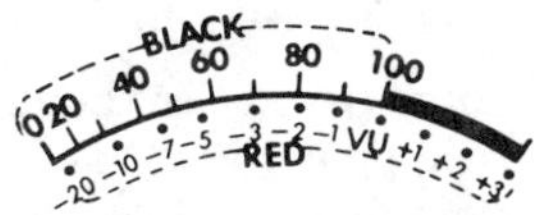

Broadcast monitoring.

Volume-indicator meter scales.

volume unit—Abbreviated VU. A unit of measurement chosen to indicate a change of 1 dB in the volume of a complex waveform. The zero reference level is a steady-state power of 1 milliwatt across an impedance of 600 ohms.

volume-unit meter—Abbreviated VU meter. *See* description under VU Meter.

vtvm—Abbreviation for vacuum-tube voltmeter.

VU—Abbreviation for volume unit.

VU meter—A meter that indicates volume on recording or playback. VU is the abbreviation for volume unit. The meter has certain special dynamic and damping characteristics.

W

W—Symbol for watt, or for work.

watt—Abbreviated W. The unit of electrical power or rate of doing electrical work. It is equal to 1 joule per second. (Mechanical power is measured in horsepower. One horsepower equals 746 watts.)

watt-hour—A unit of electrical work equal to 1 watt of power expended for a period of 1 hour. One watt-hour is equivalent to 3600 joules.

wattless power—*See* Reactive Power.

wave amplitude—The peak value of a waveform, measured from zero.

wave analyzer—An electronic test instrument capable of measuring the amplitude and frequency of the components of a complex wave.

waveform—A graphical representation of all the instantaneous values of a quantity that varies with time. All the values are connected by a line through the successive value positions on the graph.

waveform synthesizer—A device for developing a signal corresponding to a desired waveform.

wavelength—In a repeating waveform, such as a sound wave or a radio wave, the distance between any two points of corresponding phase in two consecutive cycles. The relation between wavelength (λ), frequency (f), and velocity of propagation (v) is expressed by the formula $\lambda = v/f$, where the units of distance and time are common to all terms.

wax—A medium used in mechanical recording consisting of a blend of waxes and metallic soaps.

wax master—An original recording made on a wax surface. Now obsolete as a recording medium.

wax original—*See* Wax Master.

weber—A unit of magnetic flux. One weber is the amount of flux that, when linked with a 1-turn electric circuit for 1 second, will induce an electromotive force of 1 volt in the circuit. One weber equals 10^8 maxwells.

weighted distortion factor—The weighting of harmonics in the interpretation of distortion measurement.

weighted noise level—Noise level weighted or adjusted in accordance with the 70-dB equal-loudness contour of the response of the human ear.

weighting—The adjustment of the values of factors or measurements that describe a device, according to the relative importance of each factor.

white noise—Noise in which all frequency components within the bandwidth limits are of essentially the same energy content.

whole step—*See* Whole Tone.

whole tone—Also called whole step. A major second in the equal-tempered scale. The interval between two tones having a frequency ratio approximately equal to the sixth root of 2.

wideband—Said of a tuner, amplifier, filter, etc., that responds to a broad range of frequencies.

wind—In tape recording, the way in which the tape is wound on the reel. Tape wound so that the coated surface is innermost is called an "A" wind; tape wound so that the coated surface is outermost is called a "B" wind. Windings are also classed as uniform or even, and uneven, depending on the neatness of the wind.

wire recorder—A magnetic recorder that uses a recording medium of round stainless-steel wire rather than magnetic tape.

wire recording—One of the earlier forms of magnetic recording, using a thin stainless-steel wire as the recording medium.

woofer—Common term for a speaker designed mainly to produce low audio frequencies.

work—Work is done when a force applied to a body results in a displacement component in a direction parallel to the direction of the applied force.

wow—A type of distortion in a record player or tape recorder caused by slow-speed variations in the speed of turntable or transport mechanism. (*See also* Flutter.)

wow meter—An instrument that indicates instantaneous variations in the speed of turntables or tape transport mechanisms.

wrap—1. One turn or winding of magnetic tape. 2. That part of a recording tape that is in intimate physical contact with the head at one instant.

wrap-around—The curvature displayed by magnetic tape as it passes over the pole pieces of the magnetic heads.

X — Y — Z

X—Symbol for reactance.

X_C—Symbol for capacitive reactance.

X_L—Symbol for inductive reactance.

Y—Symbol for admittance.

Z—Symbol for impedance.

zener breakdown—Reverse conduction in a semiconductor device caused by field emission of charge carriers in the depletion layer.

zener diode—A two-layer semiconductor device that exhibits a sharp knee in the reverse bias region of the current-voltage graph. When reverse voltage tends to exceed the value at this point, reverse current increases rapidly.

zener voltage—*See* Breakdown Voltage, 2.

zero beat—The condition obtained when two frequencies that are being heterodyned or beat together attain the same frequency and their beat note drops to zero.

zero bias—In a vacuum tube, circuit conditions that result in cathode and grid being at equal potentials.

zero level—Reference level against which sound or signal levels are compared. In audio-frequency measurements a zero reference of 6 milliwatts is commonly used. In some studies of sound, the threshold of hearing is used for a zero level.

zero-modulation noise—Any noise impressed on an erased tape while erase and record heads are energized for normal operation, but without an input signal.

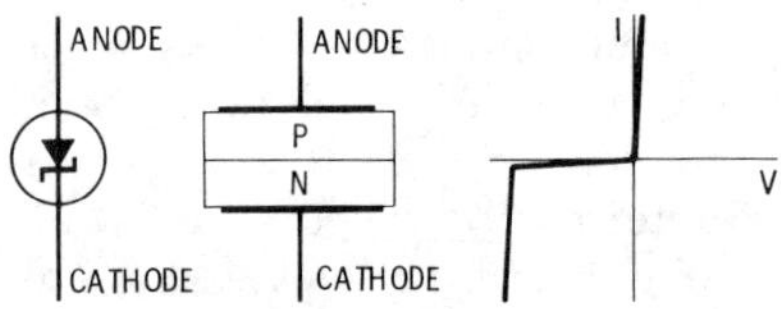

Symbol. *Junction.* *Anode characteristics.*

Zener Diode.